51 Advances in Polymer Science

Fortschritte der Hochpolymeren-Forschung

Industrial Developments

With Contributions by
G. D. Bukatov, G. Cecchin, G. Henrici-Olivé,
S. Olivé, F. A. Shutov, Y. I. Yermakov,
V. A. Zakharov, U. Zucchini

With 60 Figures and 52 Tables

Springer-Verlag
Berlin Heidelberg GmbH
1983

ISBN 978-3-662-15300-0   ISBN 978-3-540-39555-3 (eBook)
DOI 10.1007/978-3-540-39555-3

Library of Congress Catalog Card Number 61-642

This work is subject to copyright. All rights are reserved, whether the whole or part of the material is concerned, specifically those of translation, reprinting, re-use of illustrations, broadcasting, reproduction by photocopying machine or similar means, and storage in data banks. Under § 54 of the German Copyright Law where copies are made for other than private use, a fee is payable to the publisher, the amount to "Verwertungsgesellschaft Wort". Munich.

© Springer-Verlag Berlin Heidelberg 1983
Originally published by Springer-Verlag Berlin Heidelberg New York Tokyo in 1983
Softcover reprint of the hardcover 1st edition 1983

The use of general descriptive names, trademarks, etc. in this publication, even if the former are not especially identified, is not to be taken as a sign that such names, as understood by the Trade Marks and Merchandise Marks Act. may accordingly be used freely by anyone

2152/3020–543210

Editors

Prof. Hans-Joachim Cantow, Institut für Makromolekulare Chemie der Universität, Stefan-Meier-Str. 31, 7800 Freiburg i. Br., BRD

Prof. Gino Dall'Asta, SNIA VISCOSA — Centro Studi Chimico, Colleferro (Roma), Italia

Prof. Karel Dušek, Institute of Macromolecular Chemistry, Czechoslovak Academy of Sciences, 16206 Prague 616, ČSSR

Prof. John D. Ferry, Department of Chemistry, The University of Wisconsin, Madison, Wisconsin 53706, U.S.A.

Prof. Hiroshi Fujita, Department of Macromolecular Science, Osaka University, Toyonaka, Osaka, Japan

Prof. Manfred Gordon, Department of Chemistry, University of Essex, Wivenhoe Park, Colchester C 04 3 SQ, England

Dr. Gisela Henrici-Olivé, Chemical Department, University of California, San Diego, La Jolla, CA 92037, U.S.A.

Prof. Joseph P. Kennedy, Institute of Polymer Science, The University of Akron, Akron, Ohio 44325, U.S.A.

Prof. Werner Kern, Institut für Organische Chemie der Universität, 6500 Mainz, BRD

Prof. Seizo Okamura, No. 24, Minami-Goshomachi, Okazaki, Sakyo-Ku. Kyoto 606, Japan

Professor Salvador Olivé, Chemical Department, University of California, San Diego, La Jolla, CA 92037, U.S.A.

Prof. Charles G. Overberger, Department of Chemistry. The University of Michigan, Ann Arbor, Michigan 48 104, U.S.A.

Prof. Takeo Saegusa, Department of Synthetic Chemistry, Faculty of Engineering, Kyoto University, Kyoto, Japan

Prof. Günter Victor Schulz, Institut für Physikalische Chemie der Universität, 6500 Mainz, BRD

Dr. William P. Slichter, Chemical Physics Research Department, Bell Telephone Laboratories, Murray Hill, New Jersey 07971, U.S.A.

Prof. John K. Stille, Department of Chemistry. Colorado State University, Fort Collins, Colorado 80523, U.S.A.

Editorial

With the publication of Vol. 51, the editors and the publisher would like to take this opportunity to thank authors and readers for their collaboration and their efforts to meet the scientific requirements of this series. We appreciate our authors concern for the progress of Polymer Science and we also welcome the advice and critical comments of our readers.

With the publication of Vol. 51 we should also like to refer to editorial policy: *this series publishes invited, critical review articles of new developments in all areas of Polymer Science in English (authors may naturally also include works of their own).* The responsible editor, that means the editor who has invited the article, discusses the scope of the review with the author on the basis of a tentative outline which the author is asked to provide. Author and editor are responsible for the scientific quality of the contribution; the editor's name appears at the end of it.
Manuscripts must be submitted, in content, language and form satisfactory, to Springer-Verlag. Figures and formulas should be reproducible. To meet readers' wishes, the publisher adds to each volume a "volume index" which approximately characterizes the content.

Editors and publisher make all efforts to publish the manuscripts as rapidly as possible, i.e., at the maximum, six months after the submission of an accepted paper. This means that contributions from diverse areas of Polymer Science must occasionally be united in one volume. In such cases a "volume index" cannot meet all expectations, but will nevertheless provide more information than a mere volume number.

From Vol. 51 on, each volume contains a subject index.

Editors   Publisher

Table of Contents

The Chemistry of Carbon Fiber Formation from Polyacrylonitrile

G. Henrici-Olivé* and S. Olivé*
Monsanto Textiles Company, Pensacola, Florida 32575, U.S.A.

The present status of knowledge with regard to the chemical reactions and physicochemical processes, taking place during the transformation of a PAN based precursor fiber to carbon fiber, is discussed.

A large number of precursor fibers has been screened under arbitrarily chosen standard conditions of spinning, stabilization and carbonization. Included are fibers from binary copolymers, terpolymers and blends, as well as fibers containing a variety of additives.

The particular case of the AN/VBr precursor, which permits stabilization in 15–20 minutes, giving carbon fibers with excellent properties (sonic modulus close to 300 GN/m² tensile strength close to 3000 MN/m²) is described in more detail. Chemical and physicochemical reasons for the particular situation of this precursor are discussed.

* Present address:
Department of Chemistry
University of California at San Diego,
La Jolla, CA 92037, USA

1 Introduction

The first carbon fibers ever appear to have been made by Edison [1], who used them as electrical resistance in light bulbs. Prepared by pyrolysis of cellulose threads, these carbon fibers had but poor mechanical properties.

In modern times, the interest in carbon fibers is based mainly on their use as reinforcement in evpoxy or polyester resins (composites). In the early times of the development of composites, in the forties, glass fibers were used to provide tensile strength to the formable matrix, which acts as an agglomerant, and transfers stress to the fiber [2]. Glass fibers have a high tensile strength, but a relatively poor elastic modulus, so their use is confined to applications where high modulus is not required, e.g. for silos, tanks, boat bodies, etc. As increasingly more demanding end uses for composites were aspired at, e.g. by the automotive and aeronautic industry, other fibrous materials had to be developed, which would offer not only high tensile strength, but also high elastic modulus. Such fibers consist mainly of light elements such as boron, carbon or beryllium, but also of carbides, nitrides, silicides and oxides [3]. Among these, carbon fibers are potentially the most interesting (in particular for the automotive industry) because of their outstanding strength-to-weight and stiffness-to-weight ratios.

Carbon fibers consist essentially ($>99.5\%$ by weight) of carbon. They can, in principle, be made from many organic, fiber forming materials, however only three such materials have gained industrial importance: rayon, pitch and polyacrylonitrile (PAN). Rayon is injured by a relatively high carbon loss during carbonization (the oxygen contained in rayon fibers tends to be released as CO or CO_2); pitch based carbon fibers have relatively poor tensile properties, unless they are prepared from extremely purified (expensive) mesophase pitch. At present, PAN appears to be the most widely used starting material for carbon fibers.

Shindo [4] was the first to obtain carbon fibers from PAN. In the meantime, carbon fiber properties have markedly been improved, largely due to the work of Watt, Johnson and their co-workers [5] in England. The high tensile strength and modulus obtainable with carbon fibers from PAN or copolymers thereof are assumed to be due to the fact that the high molecular orientation present in stretched fibers from these materials is, at least in part, maintained through the entire pyrolysis process, provided the fibers are continuously held under stress. In the course of the process, the oriented carbon chains quasi coalesce with neighboring chains into graphitic layers, the extension of which depends mainly on the final temperature of the treatment.

On the other hand, a slow controlled thermal oligomerization of nitrile groups, in air and at moderate temperatures, leads to highly conjugated, crosslinked, oxidized structures which make the fiber non-flammable and infusible. Such fibers are said to be "stabilized." They can be further heat-treated in an inert atmosphere to form the "carbon fibers" ($>99\%$ carbon).

The stabilization process is crucial for the quality of the final carbon fiber. Fusion and other damages to the fiber have to be avoided by applying a carefully balanced regime of time and temperature gradient. For most commercial carbon fiber precursor fibers, stabilization is a time-consuming and costly procedure.

We presently report on a broad search for specific acrylic carbon fiber precursors, which should be stabilized in short time (less than one hour), and yet would give carbon fibers with satisfactory tensile properties. In planning the chemistry of such precursors, it was necessary to take into account the chemical reactions and physical processes going on during the heat treatment.

In Section 2, some properties and characteristics of PAN and acrylic copolymers, as well as of fibers spun therefrom, are discussed, with particular regard to the stabilization process.

Section 3 treats with the chemical and physical processes during stabilization and carbonization. Although literature reports have been heavily consulted, and are duely referenced, the description of the whole process, and of the relative importance of physical and chemical aspects of it, in many instances represents the authors' present opinion.

In Section 4 we report on a broad precursor screening with regard to polymer composition, additives, influence of spinning parameters, stabilization and carbonizat-ion conditions.

Section 5, finally, is dedicated to work toward the optimization of the best precursor fiber developed: fibers from copolymers of acrylonitrile with a few % of vinyl bromide. The chemical and physicochemical reasons for the particular situation of this precursor are discussed.

2 Some Characteristics of Acrylic Polymers and Fibers

Most PAN based precursor fibers are made from copolymers, the predominant component of which, however, is acrylonitrile (in most cases >95% by weight). Consequently, most physical properties of the fibers are mainly determined by this highly polar monomer.

In this Section, we review the relationship between intra- and intermolecular forces in polyacrylonitrile on the one hand, and those macroscopic properties of polymer and fiber, which are relevant to carbon fiber formation, on the other.

2.1 Dipole-Dipole Interaction of Nitrile Groups

The dominant characteristic of the PAN molecule is the presence of the strongly polar nitrile groups, at an intramolecular distance of only a few tenths of a nm. The high dipole moment (3.9 Debye) causes strong attraction or repulsion according to the orientation.

The energy involved in the interaction of two dipoles, μ_1 and μ_2, at a distance r, is generally given by (see e.g. Ref. [6]):

$$E = \frac{\mu_1 \mu_2}{r^3} (\cos \Phi - 3 \cos \delta_1 \cos \delta_2) \tag{1}$$

(angles and vectors as defined in Fig. 1). For parallel side-by-side position, $\delta_1 = \delta_2 = 90°$, $\Phi = 0$, and hence the energy has the highest attainable positive

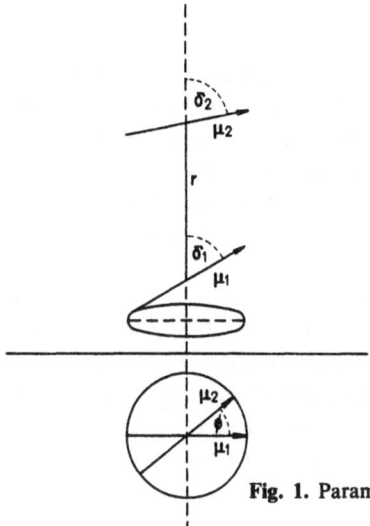

Fig. 1. Parameters involved in the interaction of two dipoles μ_1 and μ_2

value:

$$E{\uparrow\uparrow} = + \mu^2/r^3$$

indicating maximum repulsion.

The highest energy gain would be for end-to-end aligment, with

$$\delta_1 = \delta_2 = 0, \quad \Phi = 0 \quad \text{and}$$
$$E{\uparrow} = -2\,\mu^2/r^3$$

For CN groups in polymer molecules, however, this orientation is sterically not feasible. The optimum value of interaction energy is gained if the dipoles are in the antiparallel position, where $\delta_1 = 90°$, $\delta_2 = 270°$, $\Phi = 180°$, and hence:

$$E{\downarrow\uparrow} = -\mu^2/r^3$$

Low molecular weight, liquid nitriles, such as acetonitrile, are in fact highly associated, forming mainly dimers with the nitrile groups in antiparallel alignment. The energy of interaction has been estimated in the order of 20–25 kJ/mol (5–6 kcal/mol), the forces involved being mainly electrostatic in nature.

According to present knowledge, the forces effective in polyacrylonitrile, result predominantly from interactions of the strongly polar CN groups. (For a recent review of CN interactions in low molecular weight nitriles and PAN see Ref. [7]). Adjacent nitrile groups of the same macromolecule repel each other. The intramolecular repulsion compels the individual macromolecules into a somewhat irregular, helical conformation. Each macromolecule may be thought of as a more or less rigid structure, fitting within a cylinder of about 600 pm diameter. Some of the CN groups will extend beyond the confines of the cylinder. These groups are potentially available for intermolecular dipole-dipole interaction.

The CN groups assume varying angles with regard to the helical axis, guided by intramolecular repulsion and intermolecular attraction. Two adjacent macromolecules will tend to lower their energy by bringing together a maximum of groups that can interact favorably. Evidently, only a few $-C\equiv N$ groups of the macromolecules can be assumed to be oriented in such a way that perfect antiparallel alignment with $-C\equiv N$ groups of next neighbors, at minimum r, becomes possible. However, energy gain will result also from interaction of CN groups oriented in a somewhat less ideal way.

The molecular stiffness, due to the repulsion of neighboring CN groups in the same molecule is responsible for the high melting point of PAN (>300 °C). The reason is that the entropy of fusion is low, because few degrees of freedom are gained if PAN passes from the crystalline to the molten state, the intramolcular repulsion of the CN groups holding the molecule in that shiff, helical conformation.

On the other hand, the intermolecular attraction via dipole-dipole interaction makes PAN such an excellent fiber forming material. As it will be discussed later-on, the intermolecular CN interaction is also of greatest importance for the stabilization process.

2.2 Relevant Properties of Copolymers

As mentioned earlier, most work with PAN based carbon fibers in the literature has been carried out with copolymers as precursors. Initially, this may have been due mainly to the fact that most commercially available precursor fibers are of this type. However, there are also a number of other good reasons to use a copolymer. Good textile fibers generally consist of a complex system of highly ordered and amorphous zones, combining beneficial properties of each. Generally, high tensile strength, rigidity and stability of shape are anchored in the more or less crystalline regions, whereas the amorphous regions provide the fiber with elasticity and segment mobility. Stretched fibers from PAN homopolymer have excellent tensile properties, but the properties depending on the amorphous phase are not satisfactory. The reason is a relatively high degree of ordering and intermolecular CN interaction even in the "less ordered" zones. The introduction of a few percent of comonomer greatly enhances the internal mobility of polymer segments, reducing the sequence of acrylonitrile molecules capable of interacting with neighboring sequences. The most frequently used comonomers are methyl acrylate (MA) and vinyl acetate (VA).

Among the features most relevant to carbon fiber precursors, and depending on type and concentration of the comonomer, are solubility, oxygen permeation and melting point depression.

2.2.1 Solubility

The dissolution of PAN involves the decoupling of the dipole-dipole bonds between CN groups of neighboring macromolecules. Only solvents having themselves a high dipole moment, such as dimethylacetamide ($\mu = 3.81$ Debye), dimethylformamide ($\mu = 3.82$ Debye) or dimethylsulfoxide ($\mu = 3.96$ Debye) are able to dissolve PAN, and with the first of these, the solubility is so low, that the homopolymer cannot be properly spun from this solvent.

The introduction of a comonomer improves the solubility, by reducing the molecular order, and hence increasing solvent diffusion. In the case of the more common comonomers (VA, MA), about 4% (by weight) are sufficient.

Vinyl bromide (VBr) as a comonomer does not improve the solubility to the same extent. Evidently, the smaller molecular volume of CH_2=CHBr is responsible. The molecular order of PAN is less perturbed by this comonomer. As a consequence, more severe conditions are required to get a copolymer AN (96%)/VBr(4%) in solution. However, since color formation is evidently not a problem in the case of carbon fibers (color formation actually signals the onset of the stabilization reactions aimed at, cf. Sect. 3), dissolution of AN/VBr copolymers can be achieved at high temperature (110–120 °C).

2.2.2 Oxygen Permeation

The stabilization of carbon fiber precursors is done in an air atmosphere, and oxygen is required all through the fiber, for the chemical reactions going on (cf. Sect. 3).

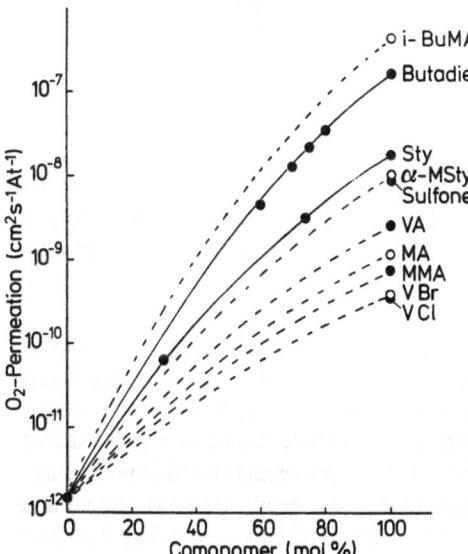

Fig. 2. Dioxygen permeation in various polymers and copolymers, at 23 °C. ●: experimental data of Salame [8] ○: calculated according to Salame [8]; VBr estimated

Oxygen diffusion in PAN homopolymer is very poor, due to the high molecular order. Again copolymerization helps to relieve the situation. Fig. 2 shows — for the cases of AN/Sty and AN/Butadiene copolymers — how the O_2-permeation increases with increasing comonomer content (experimental data of Salame [8]). The figure also includes the O_2-permeation of a number of other homopolymers (as "100% comonomer"). Assuming a similar type of dependence of the O_2-permeation on the amount of the respective monomers M in a copolymer AN/M, the O_2

permeation in such a copolymer can be estimated. The data of Fig. 2 refer to room temperature, but it may be supposed that at least the relative ease of permeation is settled by these data.

Theoretical attempts to correlate permeation with properties of the various mono- mers have not been entirely successful [8]. However, molar volume and polarity appear to be important properties, the smaller and more polar comonomers giving rise to less O_2-permeation. No data are available for vinyl bromide; however, on the grounds of molar volume and polarity, it may be assumed that O_2-permeation in VBr is similar to that in VCl.

2.2.3 Melting Point Depression

PAN homopolymer melts at approximately 320 °C [9], whereby "melting" refers to the endothermic breaking up of the highly ordered crystalline regions. The high melting point has been related to molecular stiffness (cf. Sect. 2.1).

Introduction of a comonomer into the polymeric chain depresses the melting point. A copolymer AN/VA with 7% VA, for instance, melts at $\cong 277$ °C.

The depression of the melting point of a polymer by the presence of small amounts of comonomer is described by Flory [10] and Eby [11]. Flory suggested that the como- nomer is relegated to the amorphous regions of the polymer, its presence thus reducing the extension of the crystalline regions. Flory's equation for the melting point depression caused by comonomer B in polymer A reads:

$$\frac{1}{T_m} - \frac{1}{T_m^0} = \frac{R}{\Delta H_u} X_B$$

where T_m is the melting point of the copolymer; T_m^0 is the melting point of the homopolymer A; R is the gas constant; ΔH_u is the heat of fusion per mole of crystalline repeat unit of A. Flory's equation suggests that the melting point depression depends exclusively on the molar fraction, X_B, of the comonomer, but not on its characteristics.

Actually, a linear relationship between $1/T_m$ and X_B is generally observed; however, different comonomers result in varying values for ΔH_u for the same A. Eby's theory takes care of this discrepancy, by assuming that the comonomer B is not relegated to the amorphous phase, but instead enters the crystalline lattice as a point defect. The degree to which the comonomer disrupts the intermolecular bonds that stabilize the crystalline lattice is then, to the first approximation, proportional to the volume it occupies within the lattice. An approximate form of Eby's equation has been given by Frushour [12]:

$$\frac{1}{T_m} - \frac{1}{T_m^0} = \frac{D_B}{\Delta H_u} X_B \equiv k_B X_B \tag{2}$$

where D_B is a measure of B's potential to disrupt the crystalline lattice. Since we are concerned exclusively with acrylic copolymers, and hence ΔH_u is a constant, one can use the slope k_B, of $1/T_m$ versus X_B, for a comparison of the various comonomers with regard to their ability to lower the melting point. Frushour has introduced the term "Melting Point Depression Constant" for k_B.

Table 1. Melting point depression constants, k_B, for copolymers AN/B (Frushour[12]) and molar volume of B

Comonomer B	$k_B \times 10^3$ (K^{-1})	Molar volume $(cm^3\ mol^{-1})$
Vinyl chloride (VCl)	0.51	68.7
Vinylidene chloride (VCl$_2$)	0.744	79.5
Vinyl bromide (VBr)	0.909	71.6
Ethyl vinylether (EVE)	2.2	95.0
Vinyl acetate (VA)	3.34	92.4
Methyl acrylate (MA)	3.37	90.3
Methyl methacrylate (MMA)	3.7	106.0
α-Methylstyrene (α-MSty)	4.3	129.8
i-Butyl methacrylate (i-BuMA)	4.3	160.0

Frushour has determined k_B values for some of the monomers we are presently interested in. The data are given in Table 1, together with the molar volume of the monomers under consideration. Figure 3 shows the increase of k_B with increasing molar volume, corroborating Eby's hypothesis. Deviations from linearity may be attributed to the fact that Eq. (2) is an approximation, neglecting interaction between the comonomers, such as hydrogen bonding, dipole interaction etc.

With regard to carbon fiber stabilization, Table 1 and Figure 3 clearly show that the vinyl halides depress the melting point considerably less than more voluminous monomers such as methyl acrylate, methyl methacrylate or vinyl acetate, at comparable molar level. It can be concluded that fiber fusion will be less of a problem with the former, as compared with the latter. Figure 4 illustrates this point for copolymers AN/VBr as compared to AN/VA. The melting points are calculated using Eq. (2),

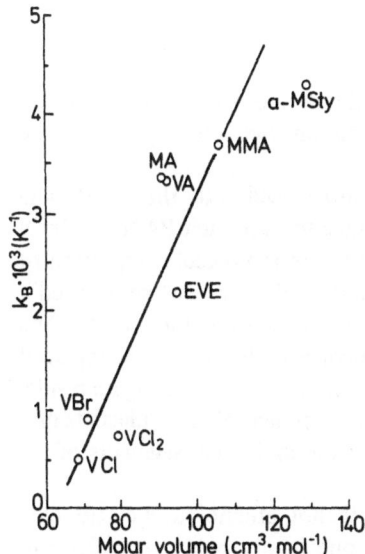

Fig. 3. Correlation between the melting point depression constant, k_B, and the molar volume of the comonomer B in AN/B copolymers (cf. Table 1)

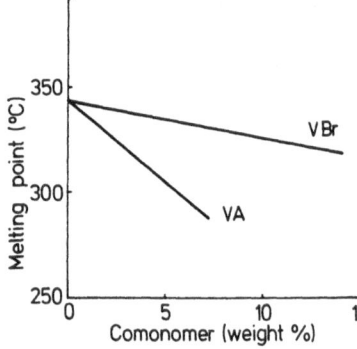

Fig. 4. Comparison of copolymers AN/VBr and AN/VA with regard to the melting point depression

with $T_m^0 = 617$ K. Molar concentrations have been converted to weight % for convenience. In the range of 7% comonomer the AN/VA copolymer melts at about 40 °C lower than the corresponding AN/VBr. Since softening of the fiber may be expected to begin some 20–30 °C below the melting point, one recognizes that fibers from a copolymer with 7% VA would fuse if stablization at about 250 °C would be intended.

At a lower temperature such fiber could certainly be fully stabilized without fusion, but this would take excessive time (5–7 hours). Such a slow thermooxidation, though effective, would be a prohibitively expensive and time consuming process.

With VBr as comonomer, the situation is different. The molar volume is very small, smaller even than that of acrylonitrile itself. It therefore does not reduce so drastically the molecular order of PAN and does not interfere so much with the intermolecular nitrile interaction. Hence it lowers the melting point considerably less. As a consequence stabilization can be carried out at a higher temperature, without danger of fusion.

2.3 The Effect of Additives

Low molecular weight substances offering a high dipole moment, such as nitriles, amides, nitro compounds, etc., have a strong plasticization effect on polyacrylonitrile fibers.

What can be expected, if such molecules are introduced into the fiber? The more mobile additive molecules have a greater chance to approach PAN — C≡N dipoles in the most favorable direction (cf. Eq. 1); the intramolecular repulsion of adjacent — C≡N groups within the polymer molecules will be partly "neutralized", with the result of less resistance to an imposed stress. At the same time, part of the intermolecular dipole-dipole interactions between polymer molecules will be replaced by interactions of polymer CN groups with dipoles of the additive, which again results in reduced tensile strength. On the other hand, the presence of the "plasticizing" compound evidently provides increased mobility for the individual segments of the polymer molecules.

Hydrogen-bonding substances, such as water, methanol, ethanol, acetic acid, etc., have a similar effect. Substances with more than one highly polar, or hydrogen-bonding group may be expected to lead to a reversible inter- and intramolecular crosslinking.

From the point of view of carbon fiber precursors, the plasticization effect of simple polar or hydrogen bonding additives translates to melting point depression, and hence is not favorable per se. However, crosslinking by such substances, having more than one binding group, may be interesting in cases of relatively low temperature heat treatments. At high temperatures (say, >250 °C) dipole-dipole bonds as well as hydrogen bonding are probably not effective.

Transition metal compounds catalyze an infinite number of organic reactions. (For a recent review, see e.g. Ref. [13]). Most probably, these compounds will have an effect on the processes going on during the heat treatment of acrylic fibers, in particular on dehydrogenation reactions, provided they can be introduced into the fiber in a way such that they will not be washed out during the spinning process.

3 The Chemistry of Stabilization and Carbonization

3.1 Oligomerization of Nitrile Groups

One of the most significant steps in the preparation of carbon fibers from acrylic precursor fibers is the oligomerization of the nitrile groups. This reaction has originally been studied in context with the problem of thermal discoloration of PAN (e.g. McCarthney [14]; Grassie and McNeill [15]; Grassie and Hay [16]). It was supposed to lead to a so-called ladder structure:

$$(3)$$

Intermolecular reaction of CN groups also takes place. The dipole-dipole interactions between CN groups of neighboring macromolecules (cf. Sect. 2.1) brings these groups into an ideal position for reaction.

$$(4)$$

The consequence of such intermolecular reaction is crosslinking:

$$(5)$$

(Concerning the relative importance of intra- and intermolecular reaction, as well as certain deviations from structures (3) and (5) in the presence of oxygen, see Sect. 3.1.3 and 3.2.1)

As all free radical polymerizations of double and triple bonds, this reaction is strongly exothermic. The heat of polymerization is of the order of 6 to 7 kcal per mol of AN monomer unit (Grassie and McGuchan [17]; Grassie [18]). The resulting structures contain conjugated $-C=N$-groups, which cause the color formation. A light yellow color can be observed at about 150 °C; after heating in vacuo at temperatures of 180–200 °C, PAN has a copper color. But the reaction becomes really important between 200 and 300 °C in air. If uncontrolled, the exothermic oligomerization of CN groups can become explosive, and the fibers fuse. However, if a suitable temperature regime and sufficient time are provided, an acrylic precursor fiber can become black, infusible and resistant to the flame. Such fiber is "stabilized." It can be further heat-treated in an inert atmosphere, to form a "carbon fiber" ($>99\%$ carbon). Heat treatment up to about 1400 °C leads to high strength carbon fiber; with heat treatment up to 3000 °C, high modulus carbon fibers ("graphite fibers") are obtained.

It should, however, be noted that a number of subsequent reactions (oxidation, dehydrogenation, condensation) take place in the later phases of stabilization; they will be discussed in detail in Section 3.2.

Most other vinylic polymers cannot be "stabilized", unless they are previously crosslinked, either chemically or by irradiation. On heat treatment, such — uncrosslinked — polymer will degrade, by main chain scission, forming radicals. In most polymers, the C—C bond of the carbon backbone is the weakest bond (cf. Table 2):

$$\sim\!\!CH_2\!-\!CH\!-\!CH_2\!-\!CH\!\sim \quad\longrightarrow\quad \sim\!\!CH_2\!-\!\overset{H}{\underset{R}{C}}^* \;+\; {}^*CH_2\!-\!CH\!\sim \tag{6}$$

Once a radical is formed, such polymer will depolymerize:

$$\underset{a}{\sim\!\!CH_2\!-\!CH\!-\!CH_2\!-\!\overset{H}{\underset{R}{C}}^*} \quad\longrightarrow\quad \underset{b}{\sim\!\!CH_2\!-\!\overset{H}{\underset{R}{C}}^* \;+\; CH_2\!=\!\underset{R}{CH}} \tag{7}$$

The reason for that is a net energy gain of 60–70 kcal/mol by transforming the radical into a double bond (Kausch [19]). This helps lowering the activation energy for the depolymerization step. As an example, the breaking of the main chain C—C bond in PVA requires $\cong 80$ kcal/mol (Table 2), whereas the activation energy for the depolymerization of this polymer is only $\cong 30$ kcal/mol [19].

Radical b has the same structure as a (Eq. 7); hence an "unzipping" of the polymeric chain may occur. In fact, many polymers such as PMMA, PMA, PSty, P-α-MSty can be depolymerized nearly quantitatively by heat treatment, with recovery of the monomer. Evidently, this depolymerization supplies "fuel" (small, volatile molecules). In the presence of O_2, and if a flame is applied or the self-ignition temperature is attained, these polymers will burn catastrophically.

Table 2. Bond dissociation energies in fragmentation of carbon chain polymers

Polymer	Bond	Dissociation energy (kJ/mol)
Polyethylene	$R-CH_2 \div CH_2-R$	$341^{[a]}$ $354^{[b]}$
	$R-CH_2 - \underset{\overset{\displaystyle \vert}{H}}{CH}-R$	$410^{[a]}$ $406^{[b]}$
Polyacrylonitrile	$R-CH_2 \div \underset{\overset{\displaystyle \vert}{CN}}{CH}-R$	$297^{[b]}$
	$R-CH_2 - \underset{\overset{\displaystyle \vert}{CN}}{CH}-R$	$450^{[b]}$
Poly(vinyl acetate)	$R-CH_2 \div \underset{\overset{\displaystyle \vert}{O-CO-CH_3}}{CH}-R$	$\cong 330^{[c]}$
	$R-CH_2 - \underset{\overset{\displaystyle \vert}{O-CO-CH_3}}{CH}-R$	$367^{[a]}$
Polysytrene	$R-CH_2 \div \underset{\overset{\displaystyle \vert}{C_6H_5}}{CH}-R$	$270^{[b]}$
	$R-CH_2 - \underset{\overset{\displaystyle \vert}{C_6H_5}}{CH}-R$	$355^{[b]}$
Poly-α-methylstyrene	$R-CH_2 \div \underset{\overset{\displaystyle \vert}{C_6H_5}}{C(CH_3)}-R$	$253^{[b]}$
	$R-CH_2 - \underset{\overset{\displaystyle \vert}{C_6H_5}}{C(CH_3)}-R$	$345^{[b]}$
	$R-CH_2 - \underset{\overset{\displaystyle \vert}{CH_3}}{C(C_6H_5)}-R$	$\cong 290^{[a]}$
Poly(methyl acrylate)	$R-CH_2 \div \underset{\overset{\displaystyle \vert}{O = COCH_3}}{CH}-R$	$\cong 340^{[c]}$
	$R-CH_2 - \underset{\overset{\displaystyle \vert}{O = COCH_3}}{CH}-R$	$373^{[a]}$
Poly(vinyl chloride)	$R-CH \div \underset{\overset{\displaystyle \vert}{Cl}}{CH}-R$	$326^{[b]}$
	$R-CH_2 - \underset{\overset{\displaystyle \vert}{Cl}}{CH}-R$	$314^{[b]}$ $334^{[a]}$

Table 2 (Continued)

Polymer	Bond	Dissociation energy (kJ/mol)
Poly(vinyl bromide)	R–CH$_2$ ⋮ CH–R 　　　　　| 　　　　　Br	330[b]
	R–CH$_2$ — CH–R 　　　　·· |·· 　　　　　Br	257[b] 284[a]

[a] Egger and Cocks [34];
[b] Stepanyan et al. [75];
[c] Estimated from similar bonds in Ref. [34]

Unstabilized fibers from PAN or acrylic copolymers also burn. However, if stabilization has been achieved, the ladder structure, as well as the crosslinking, prevent the formation of fuel, and the fibers do no longer burn.

3.1.1 Initiation of CN Oligomerization and Polymerization in Low Molecular Weight Model Substances

Evidently, the oligomerization of the nitrile groups is the clue to stabilization in PAN and related polymers. What initiates this beneficial reaction?

From work with low molecular weight organic nitriles such as acetonitrile, propionitrile, benzonitrile, and dinitriles such as maleonitrile, fumaronitrile, succinonitrile, it is known that the nitriles can be oligomerized, or even polymerized, by anionic, cationic and free radical mechanisms. (For a review, see Wöhrle [20]).

Catalysts initiating anionic oligomerization include alcoholates, alkali amides, organometallic compounds (Compagnon and Miocque [21]). In the case of aliphatic mononitriles, the strong base is assumed to abstract a labile proton from the carbon in α-position to the CN group, forming a carbanion which then attacks the CN group of a second nitrile molecule at the carbon atom (Thorpe reaction):

$$R - CH_2 - CN \xrightarrow[-HB]{+B^\ominus} R\overset{\ominus}{C}H - CN \xrightarrow{+RCH_2CN} R - \underset{\underset{CN}{|}}{CH} - \underset{\underset{CH_2-R}{|}}{C} = N^\ominus \qquad (8)$$

After this carbon-carbon bond formation, two pathways are possible, viz.
i) Formation of imines or ketones:

$$R - \underset{\underset{CN}{|}}{CH} - \underset{\underset{CH_2-R}{|}}{C} = NH \xrightarrow{+\overset{\ominus}{H_3}O} R - \underset{\underset{CN}{|}}{CH} - \underset{\underset{CH_2-R}{|}}{C} = O$$

ii) Formation of pyridine (a) or pyrimidine (b) derivatives:

a

b

c

Aromatic nitriles are trimerized to symmetric triazine derivatives (c).

The polymerization of α-ω-dinitriles is probably more relevant to the reactions taking place in PAN. Johns [22] reported that a 1:1 mixture of succinodinitrile and dimethylsulfoxide solidified within a few hours at 135 °C, in the presence of basic catalysts such as sodium methoxide, ammonia, amines, etc. Peebles and Brandrup [23] suggested the foolowing mechanism:

(9)

The cationic polymerization of mononitriles is less well understood. Kargin et al. [24] and Kabanov et al. [25], were able to polymerize several nitriles after complexing them with stoichiometric amounts of metal halides such as $TiCl_4$, $ZnCl_2$ or $SnCl_4$, by heating to 180–350 °C during several hours, in the presence of H_3PO_4 or HPO_3 as catalyst. Black polymer as well as cyclic trimers were obtained. The mechanism appears to be complex (see Wöhrle [20]).

An interesting cationic reaction takes place if α-ω-dinitriles are treated with anhydrous halogen acids. Howard [26], as well as Osborn [27], and Johnson et al. [28], studied the reaction of HBr with succinonitrile, glutaronitrile and phthalonitrile, and obtained cyclic products of the general formula

x = 1 or 2

Hydrogen bromide is considerably more active than HCl or HI. Most probably, the proton attacks at the nitrogen (Compagnon and Micque [21]; Urry [29]), e.g.:

(10)

A free radical attack upon the nitrile group was detected by Brandrup et al. [30,31]. In an investigation towards the processes responsible for chromophore formation in PAN, it was found that PAN, obtained from a normal free radical polymerization, always contains a certain amount of enamines and β-ketonitriles. Their formation was ascribed to the following reaction sequence:

(11)

Liepins et al. [32] reported the formation of black polymer from maleonitrile and fumaronitrile at 160 °C, in the presence of high temperature free radical initiators such as di-tert-butylperoxide. Based on IR and NMR, the following structure was suggested for the polymer:

Summarizing this section: The available experimental evidence points to the fact that the CN groups of PAN and its copolymers may well be expected to react in a multitude of ways under the relatively severe conditions of the "stabilization" process.

3.1.2 Initiation of CN Oligomerization in PAN and Copolymers

Polyacrylonitrile prepared by a free radical process contains generally a number of "defects", produced by the polymerization through the CN bond. This latter type of reaction is not very likely to occur with monomeric AN, because the vinylic double bond represents the highest energy, occupied molecular orbital in this molecule (Mullen and Orloff [33)]), and hence is more likely to be involved in the radical attack. In PAN, however, the highest energy molecular orbitals available are the degenerate pair of π orbitals of the CN bond. If the "wrong" polymerization has taken place with the CN group of a PAN molecule, a chain branch has been formed, and the link between the branch and the chain is a β-keto-nitrile (cf. formula 11). This group not only by itself is a chromophore (in its enol form) but, as a weak acid, it also can initiate the oligomerization of CN groups [30,31)]. Moreover, the bond that links the carbonyl group to the $-CH_2-CH-CH_2-$ chain (formula 11) is a relatively weak bond ($\cong 313$ kJ/mol; Egger and Cocks [34)]), which may break giving free radicals, which then also may be able to initiate the CN oligomerization. These reactions contribute to the slow color formation ("discoloration") in PAN, at relatively low temperatures. However, the concentration of the "defects" in the polymer is too low (about one defect per 2000 monomer units; Friedlander et al. [35)]) to account for the deep color formation and the other transformations during the stabilization process between 200 and 300 °C.

Grassie and Hay [16)] first postulated a "self-initiation" mechanism which assumed the tertiary hydrogen in PAN to be sufficiently acid to initiate a nucleophilic attack on a CN group of the second-to-next monomer unit:

However, Grassie and McGuchan [36)] shifted emphasis in favor of a free radical process. The idea of self-initiation has by now been abandoned, in particular in view of the fact that the degradation reaction (disappearance of CN groups) exhibits no isotope effect, if poly-(α-deuteroacrylonitrile) is used instead of PAN (Petcavich et al. [37)]; Grasselli et al. [38)]).

In the absence of ionic species (e.g. acidic comonomers such as itaconic acid or acrylic acid, or anionic additives such as alcoholates), the main initiation of the CN oligomerization in PAN and its copolymers is then, most probably, a radical process:

(13)

From a theoretical point of view, a homolytic main chain scission is expected to be the first consequence of heat treatment. The main chain carbon-carbon bond in polymerized acrylonitrile is the weakest bond present in PAN, as well as in those copolymers most frequently mentioned in the literature as carbon fiber precursors (AN/MA, AN/MMA, AN/VA); see Table 2.

An interesting experimental study of Balard and Meybeck [39)] makes it highly probable that free radical main chain scission in fact is the first reaction to take place. The authors synthesized oligomers of acrylonitrile:

$$R-CH(CN)-[CH_2-CH(CN)]_{n-1}-R$$

with R equal H or CH_3, and n equal 2, 3, 4, or 5. The oligomers were submitted to DTA and TGA, as well as to pyrolysis, with gaschromatographic analysis of the pyrolysis gases. The DTA showed clearly an endothermic process prior to the exothermic oligomeration of the CN groups. TGA revealed that most of the volatile degradation products are released during the endothermic phase. The eliminated gases are those to be expected from main chain scission (AN, acetonitrile, methacrylonitrile, propionitrile). Interestingly, no HCN was observed. This is in contrast to the gas evolution in PAN, and will be discussed further in Section 3.1.3.

Table 2 shows also that polymerized vinyl bromide has a C—Br bond which is even less stable than the main chain C—C bond in PAN (257 versus 297 kJ/mol, if the data of the same author are compared). Hence it may be anticipated that the CN oligomerization can be started at lower temperature with AN/VBr copolymers, with the additional advantage of no main chain scission taking place in this step.

The stabilization of carbon fiber precursors is commonly carried out in air. The oxygen has almost certainly an important effect on the initiation of CN oligomerization. (Its involvement in further steps of the stabilization process is discussed in Sect. 3.1.3 and 3.2) Once the primary radicals are formed by thermal bond breaking, a "radical multiplication" by oxygen becomes feasible. Peroxy radicals are formed which, on hydrogen abstraction (presumably a tertiary hydrogen from a neighboring molecule), create hydroperoxides as well as carbon radicals. The former decompose under the reaction conditions of stabilization; thus from one primary radical three radicals results, all of which may be able to initiate CN oligomerization:

(14)

Taking into account the exothermicity of CN oligomerization, it is evident that the reaction may not be accelerated indefinitely because undissipated reaction heat can easily lead to fiber fusion. Even a slight fusion at the fiber surface may prevent the core of the fiber from becoming fully stabilized. More severe fusion will evidently destroy the single filament structure of the fiber. A careful balance is required between — desirable — short stabilization time, and the danger of fiber fusion. An important role in this balance is played by the bulkiness as well as by the content of the comonomer, because they not only determine the ease of oxygen diffusion within the polymer, but also have a decisive influence on its melting point, and hence on its fusion behavior (Sect. 2.2). From Figs. 2–4 it may easily be concluded that a copolymer AN/VA, with a content of the relatively bulky comonomer, VA, of about 7% would be a particularly poor choice as a carbon fiber precursor. The present arguments rather point to the vinyl halides as more favorable. From the same kind of reasoning it follows that pure PAN homopolymer is not favorable either. Although the high melting point appears to be a great advantage at first sight, the extremely poor oxygen diffusion prevents a pure PAN fiber from rapid stabilization.

There are reports in the literature showing that ionic initiation of the CN oligomerization may be advantageous under certain conditions (e.g. Grassie [18]; Menikheim [40]). However, the above discussions indicates that stabilization of acrylic fibers in the absence of added ionic substances is quite feasible, provided proper attention is given to concentration as well as molar volume of the comonomer; from an economic point of view it is certainly more attractive.

3.1.3 Intra — versus Intermolecular CN Oligomerization

There is some controversy in the literature, as to whether or not isotactic placement of AN units in a polymer chain is necessary or favorable for the oligomerization reaction to take place. It has even been claimed that an increase of isotacticity would enhance somewhat the rate of oligomerization (Kubasova et al. [41]; Geiderikh et al. [42]). PAN is generally produced by a free radical process, hence the probability of occurrence of isotactic sequences is relatively low. A statistic distribution of dd, ll and dl placements (1:1:2) is more probable, and if there were any preponderance, it should be for syndiotactic sequences ldld ... rather than for isotactic lll ... or ddd ..., due to the high polarity of the CN groups which tends to hold these groups as far away from each other as possible. Moreover, atom model representations of sequences of isotactic and syndiotactic placements reveal that the former do not offer all that much of an advantage, if the two subsequent CN groups have to be "forced" into reaction distance.

The emphasis here is on the word "forced". As amply discussed in Section 2.1, the mutual repulsion between adjacent CN groups in the same chain is large, and tends to hold them apart in a helix-like structure. The dipole moment is certainly still quite high, if a free radical attack has transformed a CN group into a $R—C=N^*$ radical. Neighboring CN groups from different macromolecules, on the other hand, attract themselves. This kind of arguments appears to favor the intermolecular reaction (Formula 5) over the intramolecular reaction (Formula 3).

Fitzer et al. [43] suggested, based on the shrinkage behavior of PAN fibers during stabilization (cf. Sect. 3.3) that the presumed "ladder polymer" is in fact a "step lad-der", produced by very short sequences of intramolecular reaction (1–3 steps) followed by a jump to a CN group of a different chain. We very strongly favor this opinion. We also want to emphasize once more that the intermolecularly interacting CN groups (Formula 4) are in an ideal position for the required reaction of an activated nitrogen of one group (e.g. a radical) with the carbon atom of the other group. There is some additional experimental evidence for this opinion [44]. Differential thermal analysis (DTA) permits to observe the exothermic oligomerization reaction, under comparable conditions, for a number of different polymer compositions. Fig. 5a shows a DTA exotherm, as it is usually observed with acrylic copolymers, containing mainly AN, and a few % of one or two other monomers. (This particular copolymer is AN/VA with ca 7% VA.) Fig. 5b shows the DTA trace of an alternating AN/VCl copolymer:

$$\sim\sim \left[CH_2 - CH - CH_2 - CH \atop \qquad\ \ CN \qquad\qquad Cl \right]_x \sim\sim$$

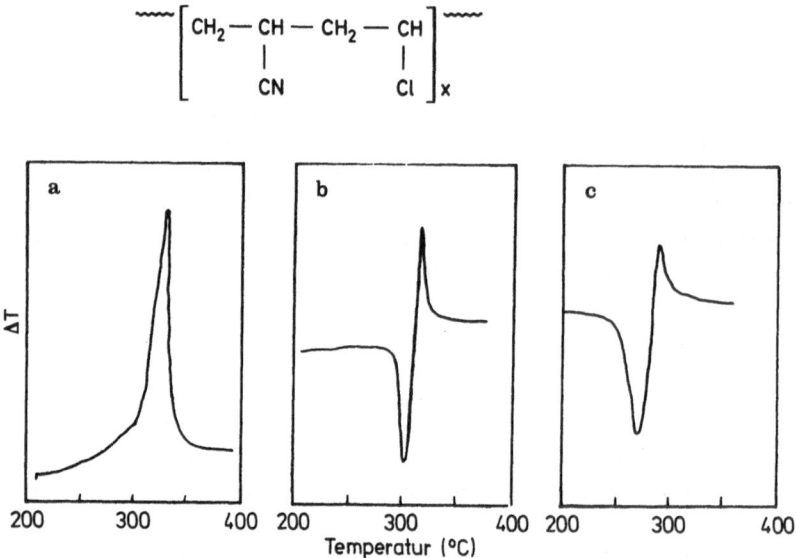

Fig. 5a–c. Demonstration of intermolecular CN oligomerization by DTA. **a** AN/vinyl acetate copolymer; **b** alternating AN/VCl copolymer; **c** statistic 1:1 AN/VCl copolymer. Exothermic peak present in **b**, where intramolecular reaction is highly improbable. (Heating rate: 20 °C/min)

(The copolymer has been described, and fully analyzed, by Wentworth and Sechrist [45] in another context). This copolymer certainly does not contain isotactic sequences of AN units. Reaction of one CN group with the next within the same macromolecule is not very probable, since it is too far away. Nevertheless, there is a clear indication of the exothermic oligomerization reaction which is, however, preceded in this case by the endothermic dehydrochlorination, typical for aliphatic chlorinated compounds:

$$\sim\sim CH_2 - CH \sim\sim \quad \xrightarrow{-HCl} \quad \sim\sim CH = CH \sim\sim \qquad\qquad (15)$$
$$\qquad\qquad\ \ \ Cl$$

The exothermic peak evidently does not account for the full amount of exothermic reaction taking place, since the exotherm is partly masked by the endotherm. Fig. 5c shows, for comparison, a statistic copolymer containing AN and VCl in an approximately 1:1 ratio (commercial product). Although sequences of AN units can be assumed to be present in this copolymer, the picture is very similar to Fig. 5b.

We conclude that the intermolecular reaction is certainly feasible, and possibly has an even greater importance than the intramolecular reaction:

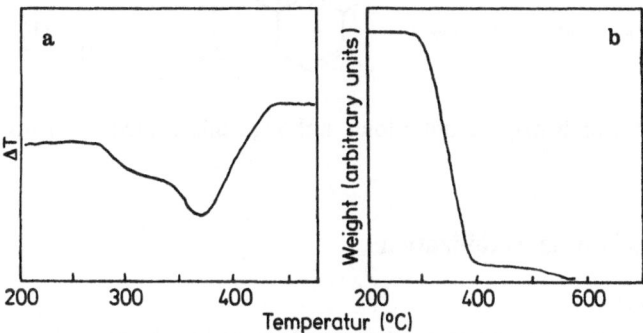

(16)

In the case of the alternating copolymer, the resulting $-C=N^*$ most probably abstracts hydrogen from any neighboring group, thus producing a new R^*. The net effect of the unification of the two CN groups is then just the formation of a crosslink. Actually, X-ray wide angle scattering of heat-treated alternating AN/VCl copolymer confirms this view: there is no evidence for the "ladder" structure, as it is usually observed with heat treated acrylic polymers (cf. Sect. 3.4.2).

In copolymers with larger sequences of AN units, there are probably one or several intramolecular steps in between crosslinks. At any rate, however, extensive crosslinking takes place already in the early stages of heat treatment.

The behavior of polymethacrylonitrile (PMAN) fits into this picture. In contrast to PAN, PMAN is not a textile fiber forming polymer, despite the presence of the highly polar CN groups. An atom model representation of a PMAN chain shows

Fig. 6a and b. DTA (a) and TGA (b) of polymethacrylonitrile; no CN oligomerization exotherm present

that the CN groups are essentially buried in between methyl groups. Hence the intermolecular dipole-dipole bonds, which are the main reason for the fiber forming capacity of PAN, are strongly hindered in PMAN. In accordance with this, the intermolecular CN oligomerization cannot take place. In fact, there is no exotherm at all to be observed [18]. Evidently the intramolecular CN oligomerization is also sterically hindered (model indicates strongly reduced rotation for the main chain C—C bonds). The only DTA feature is a broad endotherm (see Fig. 6a). Since thermogravimetric analysis (TGA, Fig. 6b) reveals a nearly complete decomposition in approximately the temperature region of the endotherm, the latter can confidently be assigned to depolymerization (and probably fusion), and it can be excluded that any appreciable CN oligomerization exotherm would be masked by the endotherm.

Summarizing this section: The intermolecular CN—CN reaction may be considered as an important — if not the most important — contribution to the exothermic oligomerization. The easy, quasi "built-in" crosslinking, upon heat treatment, is probably the most important single feature with regard to making PAN and its copolymers unique among snythetic polymers as carbon fiber precursors.

3.1.4 The Length of Cyclized Sequences

At the end of the stabilization process some 75–80% of the CN groups have reacted (IR evidence, Noh and Yu [46]). Nevertheless it is generally assumed that the individual cyclic sequences are very short. The most convincing argument is that of Fitzer et al. [43], who emphasized that the stability of condensed ring systems diminishes with increasing number of rings. The authors mentioned that, for heterocyclic rings including nitrogen, no more than five condensed rings are stable even at room temperature. Although the "jumping" from one chain to the other (cf. Formula 15) should introduce some relaxation from the strain of condensed rings, the resulting structure is still very rigid due to the continuous conjugation.

At the end of a short sequence of rings there might be a trapped radical. In fact, a build-up of paramagnetism has been observed during the heat treatment of PAN (Huron and Meybeck [47]). Alternatively, or probably simultaneously, $C=N^*$ radicals at the end of a sequence may abstract hydrogen from any surrounding CH bond, giving rise to a terminal imino group plus a carbon radical, which then would be able to initiate a new sequence of CN oligomerization.

$$+ \quad HR \quad \longrightarrow \qquad \qquad + \quad R^* \qquad (17)$$

Hence, the kinetic chain length may be quite long, although interrupted by chain transfer processes.

3.2 Further Reactions During Stabilization

3.2.1 The Effect of Oxygen

The CN oligomerization takes place in an inert atmosphere and in the presence of oxygen, although accelerated in the latter case (see e.g. Clarke and Bailey [48]). Effective stabilization in a reasonable time, however, is obtained only in the presence of oxygen (air). The following discussion of the processes taking place during stabilization refers to treatment in air.

The CN oligomerization is but the first step in a series of reactions occurring subsequently to, and partly simultaneously with, this oligomerization. The exact course of the whole process is far from being fully elucidated. The difficulty resides in the impossibility to investigate the insoluble, infusible, opaque, black reaction product with conventional analysis methods. The study of volatile decomposition products evolved during stabilization as well as during the further, carbonizing heat treatment, and infrared spectroscopy of the stabilized product have been the main sources of the thus far available information.

For the reactions occurring in immediate coinnection with the CN oligomerization, a number of reaction sequences have been suggested, based on conventional IR spectroscopy; they have been reviewed by Peebles [49]. We want to discuss here a reaction mechanism suggested by Coleman and Petcavich [50], who applied Fourier Transform Infrared (FTIR) spectroscopy to the problem, and obtained considerably more resolved spectra. Moreover, these authors observed and interpreted changes in the spectra after exchange reactions with D_2O. Coleman and Petcavich suggested a rearrangement of the cyclic structures by tautomerism, giving an imine-enamine equilibrium, followed by partial oxidation of CH_2 groups in the main chain, to give a pyridone type structure:

$$\text{(18)}$$

Although the pyridone structure had been deduced previously from conventional IR spectra by several authors, the reaction path is new, and appears to be plausible. The suggested mesomerism is based on the known fact that such mesomerism exists in tetrahydropyridine, which has an analogue structure:

In fact, in this case the enol-like form is more stable. The basic requirement for this type of rearrangement resides in a planar structure of the three atoms of the enamine group and the atoms immediately attached to this system. Under such conditions, there is a full interaction of the π-electrons of the double bond and the free electron pair on the nitrogen atom (Cook [51]). Evidently, this requirement is fulfilled in the ladder structure.

A crucial step in the rearrangement (18) is the transfer of the tertiary hydrogen to the nitrogen. Additional evidence for this step was obtained by Petcavich et al. [37] from a study of the thermal degradation of poly-α-deuteroacrylontrile using again FTIR. It could be shown that the C—D stretching frequency at 2188 cm^{-1} decreases with time upon heating (200 °C), and concurrently a broad band, centered at 2400 cm^{-1} appears, which is consistent with highly associated N—D frequencies.

The subsequent oxidation at the secondary carbon in the enamine structure (center of Formula 18) is then clearly a very favorable reaction, the CH_2 group being doubly activated by allylic double bonds:

Grasselli et al. [38], studying the kinetics of formation and disappearance of IR bands during heat treatment, found that the formation of N—H bonds is directly related to the formation of carbonyl bonds, while the disappearance of nitrile groups and formation of —C=N-groups are unrelated to N—H formation. The corollary is the following: once the enamine reacts with oxygen, enamine-imine tautomerism is no longer feasible. The stable pyridone thus determines the amount of N—H present. If, on the other hand, the N—H bonds would predominately originate from terminal imino groups —C=N—H, at the end of short sequences of oligomerized CN groups (cf. Formula 17), their intensity should be related to the disappearance of CN. Hence, this kinetic evidence corroborates the reaction sequence (18).

Coleman et al. [50] had elucidated this reaction sequence by carrying out the heat treatment under reduced oxygen pressure, in order to slow down the reaction rate, and to eliminate oxidative side reactions. Under full atmospheric pressure, they obtained a product showing a more complicated FTIR spectrum. A band at 1715 cm^{-1} has been assigned to saturated aliphatic keto groups, which can arise from oxidative attack on the polymer chain backbone, as shown by Brandrup and Peebles [30].

(19)

The latter authors pointed to the fact that such attack takes place at the secondary carbon, and not at the tertiary carbon, as one would expect from the experiences of common hydrocarbon chemistry. They explained this finding with the fact that the strongly polar CN substituent polarizes the adjacent bonds:

Hence, the elctron density is somewhat reduced at the hydrogen attached to the teriary carbon. Oxygen, being more electronegative than carbon or hydrogen, will attack at the hydrogen atom with highest electron density, i.e. at the secondary hydrogen in the main chain.

The FTIR spectrum of air-oxidized PAN reveals several more bands which have not been definitely assigned. Some authors assume the presence of OH groups, although OH bands could hardly be differentiated from the very strong NH absorptions; others believe that carboxyl groups are formed, because CO_2 is evolved in later stages of the heat treatment (see Sect. 3.4, on Carbonization). Mechanisms for the appearance of such groups have been suggested (see Review, Peebles [49]).

Prolonged heat treatment under oxygen certainly leads to dehydrogenation with concomitant aromatization of the ring structures (Huron and Meybeck [47]). So far, however, it has not been established whether a high degree of aromatization is in fact required for effective stabilization.

Thus, the oxygen appears to have a number of important tasks in the course of stabilization. On one hand, it accelerates the cyclization by the "radical multiplication process" (cf. Formula 14). On the other hand, polar groups such as $C=O$, OH and COOH, will facilitate the further condensation of ring systems. These groups (together with amino and imino groups) may also contribute to an additional initiation of CN oligomerization, via nucleophilic attack (cf. Sect. 3.1.1). It has been suggested by Watt et al. [52] that a stabilized, PAN based fiber should contain about 10% by weight of oxygen. Although much more oxygen can be taken up, the carbon yield drops markedly if higher levels of oxygen are attained.

3.2.2 Gas Evolution During Stabilization

Although the main evolution of volatiles takes place in the subsequent high temperature treatment (carbonization, Sect. 3.4), small amounts of gas are evolved during the stabilization treatment in air (>250 °C), and concomitant weight losses are reported, both increasing with the temperature (see e.g. [47]; Bell and Mulchandani [53]).

At temperatures below 240 °C, there is essentially no gas evolution, with the exception of some H_2O after long heating time. In the range 250–300 °C, small amounts of low hydrocarbons (AN, MAN, propionitrile) evolve. These compounds are supposed to proceed from depolymerization of uncyclized parts of the polymer. Hydrogen cyanide starts to evolve after 70 min at 250 °C and after 30 min at 300 °C. The early evolution of HCN has frequently been ascribed to dehydrocyanation (analog to dehydrochlorination from polyvinylchloride), either within a molecule, leading to unsaturation:

$$-CH-CH_2- \quad \longrightarrow \quad -CH=CH- \; + \; HCN$$
$$\quad\quad |$$
$$\quad CN$$

or between two chains, leading to crosslinking:

$$-CH-CH_2- \quad\quad\quad\quad -CH-CH_2-$$
$$\quad | \quad\quad\quad\quad\quad\quad\quad\quad\quad\quad |$$
$$\quad CN \quad\quad\quad\quad\quad\quad -C-CH_2- \; + \; HCN$$
$$\quad\quad\quad\quad \longrightarrow \quad\quad\quad |$$
$$\quad CH-CH_2 \quad\quad\quad\quad\quad CN$$
$$\quad |$$
$$\quad CN$$

(e.g. Watt and co-workers, s. Ref. [49]). The relatively high bond dissociation energy of the $-C \div CN$ bond (450 kJ/mol, cf. Table 2) would indicate that these reactions require higher temperature. Interestingly, Balard and Meybeck [39] did not find any HCN evolution from their oligomeric models of PAN (290 °C, air). The authors concluded that in the case of PAN, the dehydrocyanation takes place at monomer units not yet cyclized, but adjacent to cyclized and aromatized sequences. By this way the reaction would be facilitated by the gain of delocalization energy, through the extension of a conjugated system of double bonds:

$$ \text{(structure)} \longrightarrow \text{(structure)} \quad + \quad HCN \qquad (20) $$

This suggestion is corroborated by the observation that, in PAN, the hydrogen cyanide starts to evolve only after some time of heating, which would allow for cyclization and aromatization to have taken place.

Huron and Meybeck [47] did not find ammonia among the principal products of degradation of PAN in air. Earlier authors have reported the formation of NH_3, perhaps under somewhat different reaction conditions (e.g. Bell and Mulchandani [53]). The evolution of NH_3 has been interpreted as the consequence of deamination of chain ends from CN oligomerization, or of coupling of two such chain ends (Hay [54]).

3.3 Shrinkage During Stabilization

PAN based fibers develop shrinkage forces during two different stages of stabilization: entropic shrinkage in the early stages, and shrinkage due to chemical reaction in the later stages.

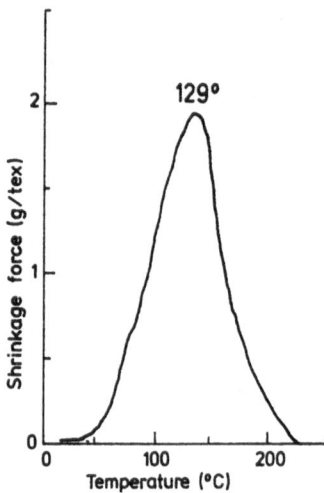

Fig. 7. Entropic shrinkage force developed by a fiber from AN/MMA (4.2%), on heating in air. Heating rate: 30 °C/min

Entropic shrinkage sets in at relatively low temperature, as soon as the T_g is reached, and hence the kinetic conditions are provided for the molecules to relax the strain acquired by stretching during the spinning operations. If a fiber is maintained at constant length, and a strain gauge is applied at one end, the force developed within the fiber during heating (the "shrinkage force") can be measured. Fig. 7 shows qualitatively the development of that force, for a fiber of AN/MMA (4.2% by weight MMA). Starting at room temperature, and heating in air at a rate of 30 °C/min., the force reaches a maximum (equivalent to some 2 g/tex), at 129 °C, i.e. at a temperature where no appreciable chemical changes occur.
the entropic shrinkage process in the first ten minutes; it will then become thermoplastic, i.e. show some elongation, and finally a second shrinkage sets in Fig. 8 shows

If a carbon fiber precursor is stabilized by holding it, at constant length, for extended time at an elevated temperature (say 3 hours at 230 °C), it will go through this in a qualitative manner. (See, e.g., Warner et al., [55]). Fitzer et al. [43] first pointed to the fact that the chemical reactions taking place during stabilization cause shrinkage of the fiber or, alternatively, if the fiber is held under a load, bring about "shrinkage force". The oligomerization of the CN groups, if jumping from one macromolecule to another ("step-ladder" principle, cf. Sect. 3.1.3), should lead to shrinkage. Segments of the polymeric chains that were parallel initially, must assume a position in which they form an angle of 120°. Fitzer estimated that shrinkages of up to 20–25% (or the corresponding shrinkage force) could result from this jumping process alone.

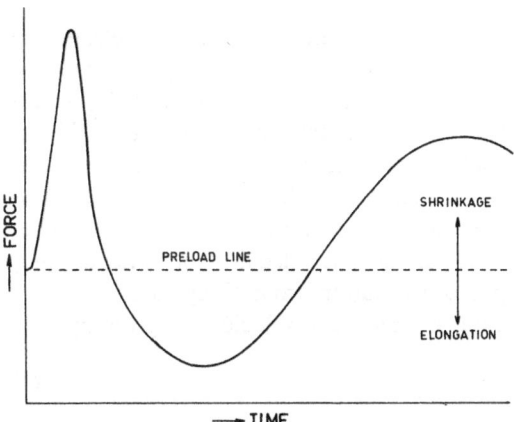

Fig. 8. Shrinkage forces developed by acrylic fibers during stabilization at constant temperature (schematic)

Qualitatively, the same sequence of shrinkage, relaxation and shrinkage (Fig. 8) is also to be expected if the fiber is stabilized dynamically, by continuously travelling through a furnace, which latter generally is not at a constant temperature, but has a characteristic profile, which may be different for different precursors (see Section 4.3.2).

It is known since time that stretching before (Prescott and Ott [56]) and/or during (Johnson et al. [5]) the stabilization treatment improves the mechanical properties of the resulting carbon fiber. Evidently, high molecular order is required. If such order has been obtained in the spinning process by extensive stretching, care must be taken

not to lose this order in the shrinkage processes. Although stretching during stabilization can take care of this task to a certain extent, the thermoplastic deformation
may cause trouble. In an unfortunate situation the fiber may respond to strain applied
from the outside (uptake rate after the furnace higher than input rate), plus the
shrinkage strain, by excessive elongation in the thermoplastic phase of its way through
the furnace; it may then tend to break in this zone. It appears that crosslinking of the
polymer, as early as possible in the whole stabilization process, could be a chemical
means to reduce this danger. Early initiation of the "step-ladder" oligomerization,
by introducing radical forming weak bonds, seems to be one possibility (VBr
comonomer, Sect. 5.1). Alternatively, one may think on a physical separation of the
various phases, by introducing rate-determining rollers in the way of the fiber through
the continuous furnace, such that the shrinkage and elongation zones are separated
from each other. Thus, entropic shrinkage could be counterbalanced effectively by
holding this zone at constant length, and tailored stretch could be applied to the
elongation phase and/or the final "chemical shrinkage" phase.

3.4 Characterization of Precursor and Stabilized Fiber

3.4.1 Orientation Factor from Sonic Modulus

High molecular orientation is generally considered to be an indispensable prerequisite
for good carbon fibers. As mentioned in Section 3.3, there is evidence in the literature
that molecular order in the precursor fiber may take effect on carbon fiber
properties.

The sound velocity in a fiber, and the sonic modulus calculated therefrom, are
related to molecular orientation (De Vries [57]). As shown by Moseley [58]), the sonic
modulus is independent of the crystallinity at temperatures well below the T_g (which
means that the inter- and intramolecular force constants controlling fiber stiffness
are not measurably different for crystalline and amorphous regions at these temperatures). An orientation parameter α, calculated from the sonic modulus, is therefore taken as a measure for the average orientation of all molecules in the sample,
regardless of the degree of crystallinity. The parameter is called the "total orientation",
as contrasted to crystalline and amorphous orientation, from X-ray data.

The relationship between sonic modulus E_s and sound velocity C is given by:

$$E_s = kC^2 \tag{21}$$

If the modulus is given in units of force per unit cross-sectional area, k is equal to the
density. If E_s is expressed in units of force per unit linear density, k is a universal
constant, depending only on the units of E_s and C. In particular, $E_s = 1.25\,C^2$, if E_s is
given in tex, and C in km/sec.

The orientation parameter α, as defined by Moseley, is:

$$\alpha = 1 - \frac{E_{s,o}}{E_s} \tag{22}$$

where $E_{s,o}$ is the sonic modulus of unoriented (randomly oriented) fiber. By
introducing $E_{s,o}$, this formulation takes account of the influence a particular

polymer structure may have on the sound velocity. However, since we are dealing exclusively with acrylic copolymers, with generally >90% of AN content, we can neglect this influence here, and use the sonic modulus of a fiber directly as a measure of molecular order therein.

Anticipating experimental results reported in detail in Section 4, we present in Fig. 9a correlation of the sonic modulus of carbon fibers and that of their precursors. Included are a variety of copolymers and blends. (For identification of the samples see Table 11.) Although there is some scattering, the trend shows clearly that high molecular order in the precursor is certainly a desirable property.

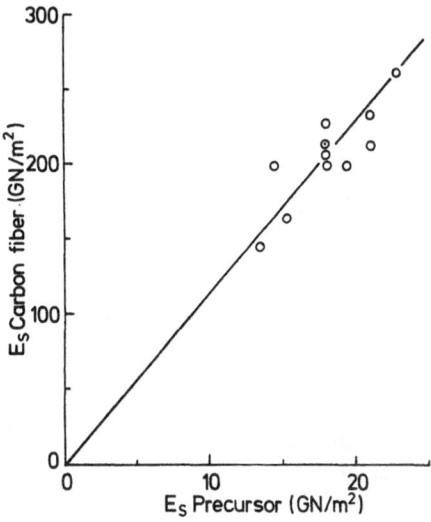

Fig. 9. Sonic modulus of carbon fibers as a function of the sonic modulus of the precursors ("standard screening conditions", cf. Table 11)

3.4.2 X-Ray Scattering Data as Criteria for Complete Stabilization

The diffraction of X-rays by crystalline matter is described by the Bragg relationship

$$\lambda = 2d \sin \theta$$

For a given wave length (generally Cu Kα radiation, with λ = 156 pm), the sinus of the angle of diffraction θ is the smaller, the larger is the separation d of the diffracting lattice planes.

The crystal structure of partly crystalline polymer, involving unit cell dimensions, repeat distances etc. in the order of a few hundreds of pm, gives rise to "wide angle" X-ray scattering. If, however, spacings of the order of 1–100 nm are concerned (e.g. dimensions of colloidal particles, of heterogeneities produced by differences in electron density, such as in fibrils, agglomerations, etc. in a polymer or fiber), very small diffraction angles result. This is the range of "small angle" scattering (e.g. d = 10 nm, 2θ = 0.9°; d = 100 nm, 2θ = 0.09°).

PAN fibers give a typical wide angle diffractogram, which undergoes characteristic changes during stabilizing heat treatment of the fiber (Uchida [59]). Fig. 10 shows

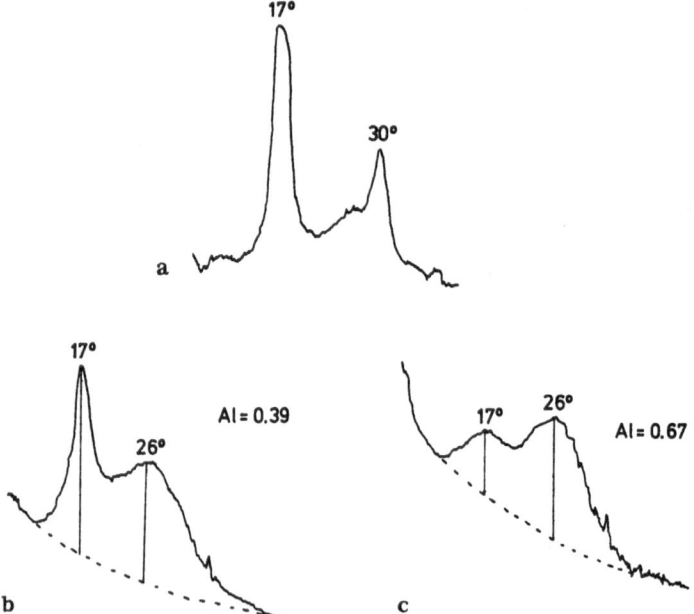

Fig. 10a–c. X-Ray wide angle diffractograms of acrylic fibers; **a** PAN untreated; **b** PAN after 60 min. 250 °C in air; **c** AN/MMA (4.2%) after 90 min. 250 °C in air; AI = aromatization index, according to Uchida [59)]

the diffractograms of untreated PAN fiber (a), of heat treated PAN fiber (b), and of heat treated AN/MMA fiber with 4.2% by weight of MMA (c). The untreated PAN shows two equatorial reflections, at Bragg angles $2\theta = 17°$ (d = 520 pm) and $2\theta = 30°$ (d = 300 pm), arising from planes more or less parallel to the fiber axis. The PAN fiber represented in Fig. 10b was heated during 60 minutes in air, at 250 °C, and was not fully stabilized (it was black, but it flashed when exposed to a match flame). The original features in the reflectogram are diminished, and a new equatorial reflection at $2\theta = 26°$ (d = 350 pm) appears. The fiber represented in Fig. 10b was heated during 90 minutes at 250 °C, and was fully stabilized.

Uchida [59)] has attributed the new reflection to the sheet-like structure of arromatized ladder polymer, and has introduced an "aromatization index", AI:

$$AI = \frac{I_A}{I_A + I_P}$$

where I_A is the intensity of the diffraction produced by the aromatized structures at $2\theta = 26°$, and I_P is the intensity at $2\theta = 17°$. The two fibers represented in Fig. 10, b and c, have IA values of 0.39 and 0.67 respectively.

Although AI values may be useful to estimate the progress of the oligomerization of the nitrile groups and the subsequent oxidation and dehydrogenation processes, they cannot be recommended as a criterion for optimum stabilization, unless a careful investigation has been made as to which IA value is actually optimum for a particular precursor. AI values close to unity may easily be obtained by prolonged heat treatment, but the ultimate carbon fiber properties deteriorate if a certain optimum oxidation time is surpassed. It is assumed that an excessive amount of $C=O$ groups in the stabilized fibers produces defects in the fiber during carbonization, by expulsion of CO or CO_2 (Bahl and Manocha [60]).

In the low angle X-ray scattering range, untreated PAN fibers have little structural detail. Well drawn fibers show an equatorial streak pattern, characteristic for fibrillar structure. Warner [61] reported a four-point pattern from a heat treated AN/MA copolymer fiber. The details of the pattern were interpreted as representing fibrils oriented along the fiber axis, with pseudocrystalline lamellar structures which are more thick than long, and oriented with a 15° declination from the fiber axis. These structures are assumed to be separated by less ordered material. Presumably the ordered structures result from some stacking of aromatized ladder polymer. Interesting as it may be from the poiht of view of a study of the incipient ordering, this kind of data does not allow a definition of maximum stabilization either.

For all practical purposes, a precursor fiber may be considered to be adequately stabilized for subsequent carbonization of this fiber displays three characteristics. First, the fiber must have maintained its mechanical integrity and retain sufficient strength after stabilization to allow for further physical handling during carbonization. Secondly, the tow bundle must not display filament-to-filament fusion due to excessive heating during stabilization. Last, the stabilized fiber must resist burning when exposed to an oxidizing flame; in this regard, the simple match test seems still to be the best.

3.5 Carbonization

3.5.1 The Chemistry of Carbonization

A fully stabilized (match test) fiber can be exposed to the high temperatures required for carbonization (1000–1600 °C). Carbonization is carried out under inert conditions (generally under nitrogen). In the course of the process, essentially all heteroatoms are expelled; carbon fiber consist to >99% of carbon. Unfortunately, the expulsion of heteroatoms does not proceed quite without loss of carbon, since volatile compounds such as HCN, CO_2, CO and CH_4 are formed. In general, roughly 50% by weight of the original PAN or copolymer ends up as carbon fiber (the carbon content in PAN is 68%). Carbonization can be achieved in short times; in a typical continuous process, at 1400 °C, a residence time in the carbonization furnace of a few minutes is generally sufficient.

Several authors have studied the volatile products evolved during carbonization, and have tried to deduce therefrom a picture of the chemistry going on. At first sight there seems to be some disagreement among the results. One should, however, take into account that different reaction conditions and experimental techniques may be involved. In particular, results of flash photolysis of unstabilized fiber or polymer are certainly not representative for the processes going on during the carbonization of the stabilized fiber. (See, e.g., Galin and Le Roy [62]). Where oxidized fiber has been treated under inert conditions, the agreement is generally good (Watt et al. [52]; Bromley [63]; Fiedler et al. [64]; Simitzis [65]).

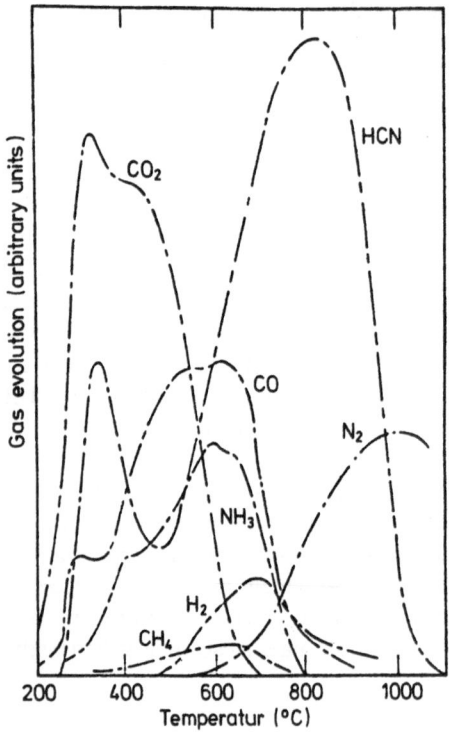

Fig. 11. Gas evolution from PAN fibers, previously oxidized in air for 5 hours at 250 °C, as a function of the temperature. Heat treatment in helium (Fiedler, Fitzer, and Rozploch [64], reprinted with permission from Carbon, *11*, 426 (1973), Pergamon Press, Ltd.).

In most cases, gas evolution has been observed as a function of temperature, and the gases have been analyzed, either by gas chromatography or by mass spectroscopy.

Fig. 11 shows representative data of Fitzer et al. [64]. The most abundant reaction product is HCN (some 15–18 % by weight, referred to the stabilized fiber, according to Bromley [63]). There are two maxima of different intensity, a small one at ca. 350 °C, and a large one at about 800 °C. It is assumed that the first results from the fragmentation of unladdered parts of the polymer (cf. Formula 20), whereas the second is the consequence of intermolecular elimination, leading to ring condensation:

CO_2 is second in abundance (7–12 %); it is evolved in the temperature range where fragmentation of unladdered polymer is assumed to be the main effect. It was suggested by Simitzis [65], that CO_2 results mainly from the decomposition of side chains of the comonomer (in this particular case MA); however Watt et al. [52] have shown that the amount of CO_2 is essentially the same for fibers from a copolymer AN/MA containing 0.4% of MA, and one containing 6%. These authors assume the presence of COOH groups, in the unladdered part of the polymer, to account for the CO_2. It should, however, be taken into account that CO, H_2 and water [63] are also found, and hence part of the CO_2 may proceed from the water gas shift equilibrium:

$$CO + H_2O \leftrightarrows CO_2 + H_2$$

and that CO elimination is the primary reaction.

NH_3 (6–8%) is assumed to be formed from imino end groups of cyclized sequences, with concomitant aromatization of a ring (Hay [54]):

CH_4 stems most probably from the fragmentation of unladdered parts, as well as from comonomer side chains. It is present only in small amounts.

Finally, N_2 starts to evolve at temperatures over 600 °C; intermolecular elimination from relatively stable, N containing, aromatic structures, with ring condensation, may safely be assumed to be the cause:

Simitzis [66] has reported elemental analyses of carbon fibers made from PAN or AN/MA copolymers, and stabilized for 3 hours at 250 °C. The fibers were carbonized for 5 minutes at different temperatures (Table 3). It follows that, under these conditions, a temperature of 1600 °C is necessary to transform the stabilized fiber to a >99% carbon fiber.

If no load is applied during carbonization, the fibers shrink also in this phase of the treatment [67]. The shrinkage of fully stabilized fiber is of the order of about 10%. It is not only a consequence of the various condensation and elimination reactions mentioned above, but also of the formation of densely packed graphitic crystallites (cf. Sect. 3.5.2). The shrinkage can, however, be counteracted by a load, and the mechanical properties of the carbon fiber are improved if this is done.

Table 3. Elementary analyses of carbon fibers [66]. Stabilization: 3 h, 250 °C; time of carbonization: 5 min

Carbonization Temperature	Element (weight %)	PAN	AN/MA (1%)	AN/MA (5%)
1000 °C	C	90.71	90.38	89.36
	H	0.89	0.85	0.84
	N	6.60	7.00	7.80
	O	1.80	1.77	2.00
1200 °C	C	96.45	96.40	96.20
	H	0.25	0.20	0.20
	N	3.30	3.40	3.60
	O	—	—	—
1400 °C	C	97.60	97.19	97.10
	H	0.10	0.12	0.15
	N	2.30	2.69	2.75
	O	—	—	—
1600 °C	C	$\gtrsim 99.5$	$\gtrsim 99.5$	$\gtrsim 99.5$
	H	—	—	—
	N	$\gtrsim 0.5$	$\gtrsim 0.5$	$\gtrsim 0.5$
	O	—	—	—

3.5.2 Structure and Properties of Carbon Fibers

Bennett and Johnson [68] examined, by electron microscopy, longitudinal and transversal sections of PAN based carbon fibers. The authors also reviewed and criticized previous work.

Lattice fringe images, produced by phase contrast methods, appear to have given the most reliable information. Electron micrographs of species heat treated at 300–800 °C revealed developing order, length and packing of the lattice fringes. Stacking sizes of incipient layer planes from 1.2 nm at 410 °C to about 1.7 nm at 800 °C were measured. Taken in conjunction with the chemical analyses of the volatile products, these results indicate an interlinking of ladder polymer sequences along the direction of the incipient crystallites present in stabilized fiber. The lengthwise interlinking is assumed to be more favorable for crystallites aligned in the direction of the fiber axis. From the electron microscopic study of high modulus carbon fiber ("graphite fiber", heat treated at 2500 °C), Crawford [69] concluded that these fibers consist of graphitic crystallites, the layer planes of which interlink in a highly complex manner, giving rise to a range of crystallite sizes, and enclosing extensive voids. The layer planes are predominantly oriented along the fiber axis, but there is no discrete fibrillar structure.

Although the typical high tensile strength carbon fibers (obtained at heat treatment < 1600 °C) are certainly considerably less ordered, and less densely packed, their structure may be assumed to be somewhere in-between the fibrillar structure with some pseudocrystalline lamellar order within the fibrils, as suggested by Warner [61] for stabilized fiber, and the graphitic crystallites suggested by Crawford [69]. The voids determine the relatively low density of the carbon fibers (for PAN based fibers

1.6–1.9 g/ml, according to heat treatment), as compared with the theoretical density of graphite (2.26 g/ml).

The graphitic layer planes are held together only by weak van der Waals forces, whereas strong covalent sp^2 bonding is present within the layers. The elastic modulus of the fiber is assumed to be determined mainly by the size, and the orientation in the fiber axis, of these graphitic layers, which both increase with increasing temperature of carbonization. Carbon fibers (<1600 °C) have modulus values in the order of 200–300 GN/m², whereas "graphite fibers" may attain up to ca 700 GN/m² (Diefendorf and Tokarsky [70]).

The tensile strength is, theoretically, related to the modulus by the — simplified — Griffith equation for brittle fracture:

$$\sigma = \left(\frac{2E\gamma_e}{\pi c} \right)^{1/2} \tag{23}$$

where σ = tensile strength at break
       E = elastic modulus
       λ_e = specific elastic surface energy
       c = critical crack depth

However, other factors, particularly the presence of flaws, may limit the observed strength. From selected PAN based carbon fibers, Diefendorf and Tokarsky determined an experimental average value of $(2\gamma_e/\pi c)^{1/2} = 4650 \text{ N}^{1/2}/\text{m}$ (= 56 psi$^{1/2}$). Hence, the relationship between modulus and tensile strength at break, for PAN based fibers, may be expected to be

$$\sigma = 4650E^{1/2} \tag{24}$$

as long as no flaws have been introduced in the course of spinning, stabilization and carbonization operations.

At a heat treatment above 1600 °C, the tensile strength is known to fall below the value expected from Eq. 24. In fact, σ versus temperature curves show a clear

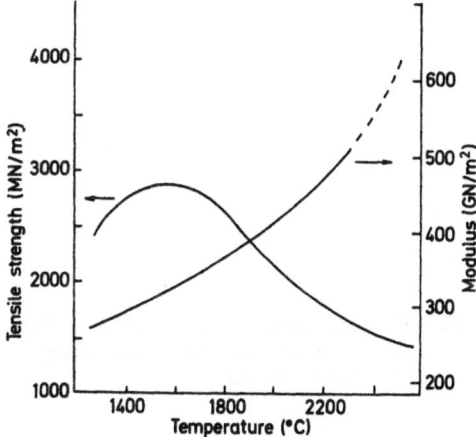

Fig. 12. Approximate dependence of modulus and tensile strength on carbonization temperature for PAN based carbon fibers

maximum for this temperature range, that may come close to 3000 MN/m², to decrease then to values below 1500 MN/m² for the high temperature graphite fibers [70]. It appears that a more and more ordered skin is formed, comprising larger and more perfectly oriented crystallites than the core [68]. Diefendorf and Tokarsky [70] assume that a microcompressive buckling of the ordered skin occurs on cool-down from the final heat treatment, generating microcracks. Moreover, extensive crystallization at the fiber surface is thought to produce volume flaws which limit the strength.

The approximate relationship between modulus and tensile strength of PAN based carbon fibers, as a function of the carbonization temperature, as it may be expected for ideal cases according to the mentioned literature, is given in Fig. 12.

4 Discussion of Specific PAN Based Precursors

4.1 The Importance of Type and Concentration of Comonomer

As mentioned earlier, most work with PAN based carbon fibers in the literature is done with copolymers as precursors, and there are several good reasons for that.

Figure 13 shows DTA curves of pure PAN (a), of a copolymer AN/VA with 7.4% VA (b), and AN/VBr with 6.4% VBr (c). Pure PAN gives a very sharp, narrow exotherm, indicating that the exothermic reaction, i.e., the oligomerization of the CN groups, takes place violently, in a very short time and small temperature

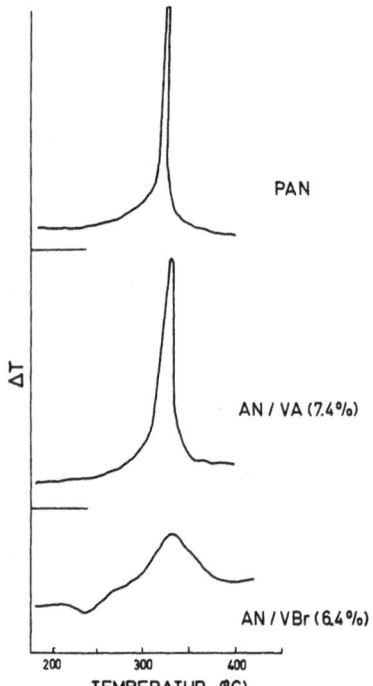

Fig. 13. Differential thermal analysis (DTA) of PAN and copolymers. Heating rate: 20 °C/min. (scale of ordinate ca 25 × that of abscissa)

interval. It can be estimated that, under the experimental conditions of the DTA measurement, most of the oligomerization occurs within one minute, as soon as the temperature is attained where radical breaking of the weakest bonds (main chain scission, cf. Table 2) provides sufficient initiating radicals, and segment movement permits the oligomerization process to proceed. Figures 13b and c show that the exothermic maximum is flattened and the heat release is spread over a broader temperature range, if a comonomer is present. The effect is considerably more marked with the VBr copolymer than with AN/VA, although somewhat less comonomer is present in the former case.

Admittedly, the experimental conditions of DTA (heating rate of 20 °C per minute) are not strictly representative for the conditions during slow stabilization. Thus, it is known that the form of the DTA curve, as well as the temperature of the maximum may change considerably with the heating rate. Nevertheless, data obtained under identical conditions certainly indicate differences between the three samples, which are very relevant to the stabilization process. Grassie et al. [18] have investigated a very large number of copolymers by DTA. The broadening of the exotherm was found always, whatever comonomer was chosen, and even at concentrations in the range of 1–2% (although, of course, more marked at higher levels of comonomer). Certain comonomers (e.g. itaconic acid or acrylic acid) are assumed to start, via an ionic initiation, the oligomerization of CN groups at a temperature lower than that required for main chain scission, and thus to help to spread the exotherm over a broader temperature range [18]. As it will be shown in Section 5.1, VBr belongs also in this category, although the additional initiation is mainly of the free radical type.

For the more inert monomers, such as MMA, MA, VA, it does not appear very probable that they are directly involved in the cyclization reaction, in particular not at the low concentration level they are generally used in carbon fiber precursors. We suggest that their influence is more of a physical than of a chemical nature. On the one hand, the presence of a comonomer will increase the chain segment mobility (cf. Sect. 2.2), and the oxygen permeation through the fiber (Sect. 2.2.2). This will certainly enhance chain initiation at lower temperature. On the other hand, the strong intermolecular dipole-dipole interaction is reduced, even by small amounts of comonomer, as evidenced by the improvement of the solubility (Sect. 2.2.1) and by the lowering of the T_g and the melting point (Sect. 2.2.3). Since intermolecular dipole-dipole interaction between CN groups should be considered as a very important — if not the most important — contribution to the oligomerization reaction (Sect. 3.1.3), a reduction of the rate of chain propagation is to be expected. Hence, the introduction of a comonomer will spread the oligomerization over a broad temperature range or — at constant temperature — over a larger time interval, thus helping to dissipate the large reaction heat.

However, a price has to be paid: there is a concomitant melting point depression which is detrimental to the stabilization operation. As an example, it had been shown in Fig. 4 that the melting point of AN/VA ($\simeq 7\%$ VA) is reduced by some 50 °C compared to PAN. Since even a slight superficial fusion may prevent the fiber from becoming fully stabilized, the thermal treatment of such a fiber has to start at a relatively low temperature (say, 220 °C). Only after a certain amount of cross linking has taken place, the fiber may be gradually heated to a higher temperature

Table 4. Static stabilization at constant length; comparison of different copolymers (250 °C, 1 hour)

Copolymer	Comonomer (weight %)	Fusion	Stabilization
PAN	—	not fused	—
AN/VA	3.1	not fused	—
	3.9	not fused	+
	4.2	fused	—
	7.4	badly fused	—
AN/MA	2.9	not fused	—
	3.9	slightly fused	±
	5.3	fused	—
AN/MMA	3.9	not fused	—
	4.2	not fused	+
	4.4	slightly fused	—
AN/VBr	4.2	not fused	—
	6.4	not fused	—
	14.0	slightly fused	—

(say, 300 °C) to speed up the stabilization process, but even so some 5–7 hours are required for the stabilization of fibers from AN/VA with $\cong 7\%$ VA.

To stabilize PAN based fibers in a reasonable time (say one hour or less), higher entrance temperatures are required, and this inevitably dictates limitations concerning the type and the concentration of the comonomer. Evidently, a compromise between the beneficial influences of the comonomer with regard to the oligomerization reaction, and the danger of fiber fusion, has to be looked for.

As discussed in Section 2.2.3, the melting point depression is the more pronounced, the larger the concentration of the comonomer and the larger its molar volume. The more commonly used comonomers, viz, MMA, MA and VA, are similar with regard to their specific influence on the melting point depression (cf. Fig. 3). Table 4 shows results of static stabilization at constant length. The fibers were wrapped on a glass frame and heated in a laboratory oven at 250 °C, for 60 minutes. The Table indicates that the copolymer fibers do not, or only very slightly, fuse under these conditions, as long as they contain ca 4% or less of these comonomers. At ca 4%, the fibers are either fully stabilized, or very nearly so. At lower concentration, as well as in pure PAN, the fibers do not fuse but they are not stabilized. Evidently, the high molecular order in these fibers does not permit sufficient segment mobility and oxygen diffusion. It follows that the three copolymers under consideration behave very similarly with regard to fusion and stabilization, and can be stabilized in relatively short times, if the correct concentration level is chosen for a given temperature. In the particular case of AN/MA, AN/MMA and AN/VA, this happens to be ca 4% by weight, for the experimental conditions used.

Fibers from AN/VBr copolymers, which are also included in Table 4, behave differently, as it was to be expected from the data exposed in Figs. 3 and 4. Due to its small molar volume, VBr depresses the melting point only slightly, even at relatively high concentrations. Thus, under the conditions of Table 4, the fibers with 6.4% VBr content do not fuse, and even with 14% there is only very slight fusion. But the fibers are not stabilized. The small VBr molecules do not disturb essentially the high molecular order in PAN, and consequently O_2 diffusion and segment mobility are low. However, these fibers can tolerate considerably higher entrance temperatures, and behave very favorably under such conditions, as it will be seen in the following sections.

4.2 Precursor Screening

Numerous copolymer compositions have been screened with regard to ultimate carbon fiber properties, under comparable conditions of spinning, stabilization and carbonization. Although it is evident that different compositions will require different conditions for optimal properties, it was felt that a standard screening procedure, taking care of complete stabilization (match test), and providing for a minimum of fiber breaks, should help to select potential candidates.

The "Standard Screening Conditions" comprised:

a) Spinning at a sixfold overall stretch.

b) Stabilization in a continuous way, in a furnace of 5.7 m length, without stretching (i.e. at constant fiber length), with a temperature profile as shown in Fig. 14, and a residence time of 60 min.

c) Carbonization in a continuous way, in a furnace of 75 cm length, with a residence time of 6.7 min, at 1400 °C.

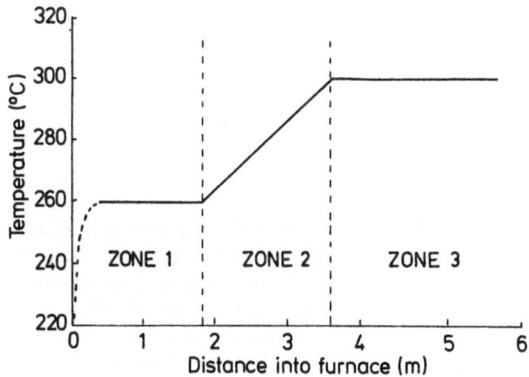

Fig. 14. Temperature profile in three-zone stabilization furnace ("standard screening conditions")

The quality of the carbon fibers was evaluated mainly from the tensile properties, modulus E and tensile strength σ. Sonic modulus E_s was used instead of tensile modulus, because the method appeared to be more reliable. Apart from E_s and σ, the carbon fiber density ϱ was determined.

4.2.1 Binary Copolymers

Table 5 shows the results from screening of fibers spun from binary copolymers as precursors. The fibers spun from AN/VBr copolymers are clearly the best with regard to carbon fiber properties.

Table 5. Precursor screening: binary copolymers

Precursor AN/M		Carbon Fiber Properties		
Comonomer M	Weight % M	E_s (GN/m²)	σ (MN/m²)	ϱ (g/cm³)
VBr	4.2	234	2200	1.64
VBr	4.2	214	2090	1.67
VBr	5.2	234	2160	1.64
VBr	6.4	214	2000	1.61
MMA	4.4	214	1500	1.59
MMA	4.2	207	1500	1.65
MMA	3.2	200	1490	1.68
MAN[a]	5.0	200	1400	1.50
MA	3.9	200	1380	1.68
VA	3.1	117.	930	1.50
VA	3.9	165	1200	1.66
VA	4.0	159	1070	1.54
AAM[b]	4.0	200	1190	1.59

[a] MAN = methacrylonitrile; [b] AAM = acrylamide

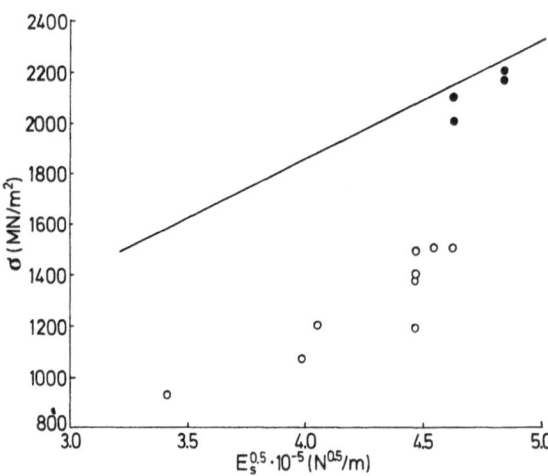

Fig. 15. Carbon fibers from binary copolymers, prepared under "standard screening conditions" (cf. Table 5), ●: AN/VBr precursor; ○: all others; solid line calculated with Eq. (24)

Fig. 15 checks these screening data against the empirical Eq. (24). It appears significant that, with the exception of the first four AN/VBr fibers which are close to the calculated line, all carbon fibers do not have the tensile strength that would correspond to their respective moduli. This fact suggests excessive flaws, that might have been introduced either by spinning, or during the stabilization process. Since the empirical Eq. (24) was obtained from commercial precursors, that almost certainly had been stabilized under milder conditions (longer stabilization times), we tend to assume that the standard screening stabilization conditions chosen (aiming at short stabilization time), are too rough for most of these copolymers. In particular it appears that the furnace entrance temperature is too high for all but AN/VBr (see Sect. 4.3.2).

4.2.2 Terpolymers and Blends

A number of terpolymers (Table 6) and of blends of copolymers (Table 7) have been investigated in an attempt to pool beneficial properties of two different comonomers. Thus, VBr was combined with MMA, MA or VA to improve the oxygen permeation (cf. Sect. 2.2.2). α-Methylstyrene (αMSty), as well as acrolein, were used

Table 6. Precursor screening: terpolymers AN/A/B

Precursor Composition		Carbon Fiber Properties		
A (weight %)	B (weight %)	E_s (GN/m^2)	σ (MN/m^2)	ϱ (g/cm^3)
MA (3.5)	VBr (5.0)	214	1780	1.57
MMA (1.6)	α-MSty (1.0)	200	1730	1.59
MMA (2.0)	Acrolein (2.0)	200	1310	1.63
VA (2.2)	IA (3.9)	200	1270	1.60
MA (1.8)	α-MSty (2.0)	158	1180	1.60
MMA (3.0)	NMAA (1.0)	200	1240	1.64

Table 7. Precursor screening: blends of copolymers

Precursor Composition			Carbon Fiber Properties		
Copolymers	Comonomer (weight %)	blend ratio	E_s (GN/m^2)	σ (MN/m^2)	ϱ (g/cm^3)
AN/MNA AN/VBr	4.4 4.2	1:1	213	2060	1.64
AN/MAA AN/VBr AN/MMA	3.2 14.1 4.0	5:3:2	200	1850	1.62
AN/VA AN/VBr	4.0 4.3	1:1	180	1630	1.64
AN/VA AN/HEMA	4.0 20.0	9:1	158	1330	1.54

to introduce weak bonds able to give free radicals at relatively low temperatures (cf. Table 2); itaconic acid (IA) and methacrylic acid (MAA) were expected to give additional ionic initiation of CN oligomerization; N-methylolacrylamide (NMAA) is a crosslinking agent that becomes effective at relatively low temperatures. Hydroxyethyl methacrylate (HEMA) is strongly hydrogen-bond forming.

Although in some of the cases moderate improvements over the simple binary compositions may be discerned from the Tables 6 and 7, none of the compositions was superior to the binary AN/VBr copolymers, at least not under the screening conditions. It should be noted that VBr containing compositions are again at the top of each of the Tables. The 1:1 blend of AN/MMA + AN/VBr (first in Table 7) is particularly interesting because it shows carbon fiber properties comparable to the binary compositions AN/VBr (4–6% VBr) with an overall content of VBr of only 2%.

4.2.3 The Effect of Crosslinking Agents

Crosslinking by means of multifunctional compounds, either by coordinative bonding (hydrogen bonding, dipole-dipole interaction, cf. Sect. 2.3) or with chemical bond formation, was investigated. It was expected that crosslinking in the very early stages of stabilization would reduce the danger of fiber fusion, and also counteract molecular fracture, thus reducing carbon losses and/or improving carbon fiber tensile properties.

Compounds like dimeric sugars (e.g. sucrose) or polyacids (e.g. H_3BO_3), are of the hydrogen-bonding type. In the case of fibers from AN/VA (3.1%), the addition of H_3BO_3 or sucrose to the polymer, prior to spinning, improved the carbon fiber properties substantially; fibers from AN/VBr, however, did not show this positive response (see Table 8).

Table 8. Influence of crosslinking additives (standard screening conditions)

Precursor Composition		Carbon Fiber Properties	
Copolymer[a]	Additive[b]	E_s (GN/m²)	σ (MN/m²)
AN/VA (3.1%)	—	117	930
	H_3BO_3 (4%)	200	1000
	Sucrose (4%)	165	1400
AN/VBr (4.2%)	—	213	2090
	Sucrose (4%)	213	1520
AN/MMA (3.2%)	—	200	1490
	Maleic Anhydride (0.5%)	179	1560

[a] In parentheses weight % of comonomer;
[b] in parentheses weight % of additive, referred to overall polymer

Crosslinking by chemical bond formation was tried with maleic anhydride. It is known that the addition of dienophiles such as maleic anhydride or acrylamide reduces color formation ("discoloration") in acrylic fibers. A reasonable explanation for this phenomenon has been given by Marien [71]. It is based on a Diels-Alder reaction of the dienophile with cis-dienic structures created during the oligomerization of the nitrile groups:

Polyacrylonitrile → Polyimines

Dienophile

Non-conjugated
imine Copolymer

(Note that such cis-dienic structures can be present only if intermolecular oligomerization of nitrile groups has taken place.) The Diels-Alder product interrupts conjugation, thus reducing color formation. In the particular case of maleic anhydride, and
under the relatively rough conditions of stabilization, this type of reaction should
lead to crosslinking, by cross-anhydridization:

A polymer solution containing AN/MMA copolymer plus 5% (referred to polymer)
of maleic anhydride gelled completely, indicating that the expected crosslinking
actually took place. As the maleic anhydride concentration was reduced to 0.5%,
fiber spinning was no longer a problem. However, at this low concentration the
beneficial effect on carbon fiber properties was only minor (see Table 8).

The question of reduced stabilization time has not been investigated, since the
"standard" conditions have been applied throughout. Only in one case it was
observed that a fiber of AN/MMA (3.2%) with 0.5% of maleic anhydride was fully
stabilized in one hour at 250 °C, whereas the parent fiber (no additive) was not,
under the same conditions.

4.3 Influence of Spinning and Stabilization Parameters

4.3.1 Stretching

Whereas the screening of precursors had been made purposely at a constant set of conditions, it was clear that this was not necessarily the optimum. The advantage of stretching during spinning and stabilization is known from the literature (cf. Sect. 3.3).

Table 9. Influence of stretching on carbon fiber properties

Precursor Composition[a]	Stretching		Carbon Fiber Properties		
	Spinning (X)	Stabilization (%)	E_s (GN/m²)	σ (MN/m²)	ϱ (g/cm³)
AN/MMA	6	0	207	1510	1.65
	10	0	213	1760	1.67
AN/VBr	6	0	213	2090	1.67
	13	0	262	2270	1.70
AN/VA + AN/VBr	6	0	179	1630	1.64
	6	15	213	1650	1.63
AN/VBr	6	0	213	1800	1.59
	6	15	234	2070	1.59
AN/VBr	13	0	262	2270	1.70
	13	5	262	2480	1.72
	13	10	262	2200	1.72

[a] Content of comonomer M in all copolymers AN/M: 4.2 weight %

Table 9 shows, in the upper part, two examples where the spin stretch has been varied, all other parameters being held constant. The improvement of carbon fiber properties is evident. In the lower part, the Table shows 5 examples of broadly varying precursor fibers which, maintaining again all other conditions, including spin stretch, constant, have been stabilized without and with stretching each. In general, the response with regard to carbon fiber properties is positive. But evidently there is an upper limit above which the fiber is damaged. The amount of stretching a fiber can favorably take during stabilization, depends not only on the particular precursor, but also on the stretching given to it during spinning. The last example in Table 9 illustrates this: for an AN/VBr fiber with a 13-fold spinning stretch, 10% stretching during stabilization is already damaging.

It appears then evident that for each precursor composition the stretching during spinning and stabilization has to be carefully fine-tuned.

4.3.2 Stabilization Profile (Temperature/Time Regime)

The time a fiber spends in the stabilization furnace, as well as the temperature profile it sees during its traveling through the furnace, are of the uttermost importance for

the ultimate carbon fiber properties. The dangers involved are:
— incomplete stabilization
— fiber fusion
— overoxidation.

The response of different precursor fibers to different temperature/time regimes shows dramatic variations. Fig. 16 shows the profiles discussed in this Section. Profile A corresponds to the "standard stabilization conditions", cf. Fig. 14.

The different response of varying precursor fibers is exemplified in Table 10. In each group of experiments, two carbon fibers are compared, which had been obtained from the same precursor, under identical conditions, except for different profiles during stabilization.

The first two groups compare a slow, mild stabilization (profile C) with a short, rather rough stabilization (profile A). Whereas the terpolymer AN/MA/VBr clearly

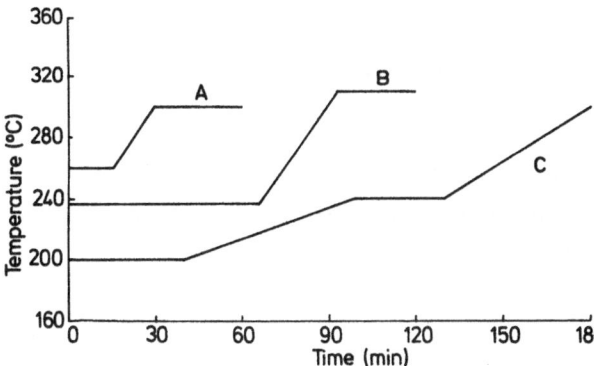

Fig. 16. Stabilization temperature/time profiles discussed in Sect. 4.3.2

Table 10. Influence of stabilization profile (temperature/time regime) on carbon fiber properties. Profile definition see Fig. 16; no stretching in stabilization·

Precursor	Stabilization	Carbon Fiber Properties		
Composition[a]	Profile	E_s (GN/m²)	σ (MN/m²)	ϱ (g/cm³)
AN/MA (3.5%)/VBr (5%)	C	228	2190	1.67
	A	214	1780	1.57
AN/VBr (6.4%)	C	165	1730	1.64
	A	214	2000	1.61
AN/MMA (4.2%)	A	207	1500	1.65
	B	213	2000	1.52
AN/VA (3.9%)	A	165	1200	1.66
	B	199	1700	1.52
AN/VBr (4.2%)	A	214	2090	1.67
	B	214	1940	1.61
AN/MMA (3.2%) + 0.5% maleic anhydride	A	179	1560	1.63
	B	214	1660	1.58

[a] Comonomers in weight %

benefits from the milder conditions, the binary composition AN/VBr appears to require the higher temperature of profile A.

The remaining four groups compare profiles A and B, whereby B is intermediate between A and C with regard to entrance temperature and time, but has a somewhat higher end temperature (see Fig. 16). Profile B turns out to be better in all cases, except for AN/VBr.

It appears evident from the results of Table 10 that each precursor composition requires individual optimization with regard to the stabilization temperature/time regime. Fibers based on AN/MMA copolymer for instance, are favored by lower entrance temperature, and probably do not tolerate well high end temperature. Stabilization at a constant (or only slightly increasing) intermediate temperature appears to be the most promising approach. AN/VBr based fibers, on the other hand, tolerate remarkably high entrance temperature, and require high final temperature for full stabilization. This observation clearly points to AN/VBr as the best candidate for short stabilization times. Nevertheless, all other precursors have probably potential for improvement by optimization of the stabilization conditions.

4.4 Correlation between Presursor and Carbon Fiber Properties

4.4.1 Molecular Orientation

The importance of molecular orientation has been discussed in Section 3.4.1. In particular, the linear relationship between the sonic modulus of the carbon fiber and that of the precursor was exemplified in Fig. 9.

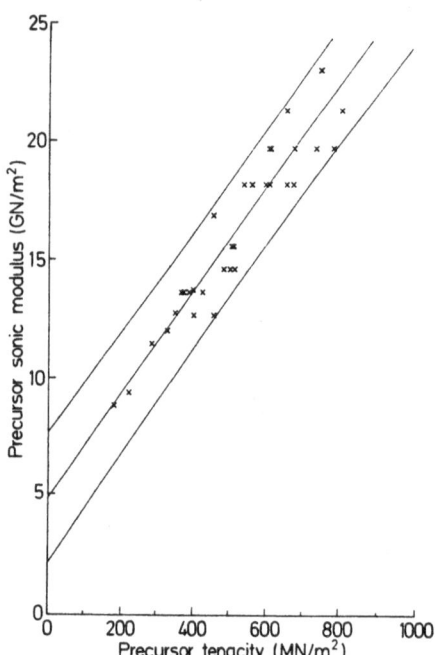

Fig. 17. Modulus/tenacity relationship for precursor fibers (computer output for extrapolation to tenacity = 0; data see Table 11)

Fig. 17 shows that the single filament tenacity (Instron) of the precursor may be considered as a suitable, simple indicator for molecular orientation. For a wide variety of precursor compositions, comonomer concentrations and spinning conditions, there is a relatively good linear relationship between sonic modulus and tenacity of the precursor fiber. The identification of the data points is given in Table 11. (Not all of the fibers used for Fig. 17 have been actually transformed to carbon fibers; some of the AN/VBr carbon fiber data are taken from Sect. 5.)

It appears plausible to identify the extrapolated value of the modulus at tenacity zero as the sonic modulus of "unoriented fiber", $E_{s,o}$ (cf. Eq. 22). This magnitude,

Table 11. Sonic modulus and tenacity of precursor fibers, and sonic modulus of carbon fibers (where determined under "standard stabilization conditions"; cf. Fig. 9).

Precursor				Carbon Fiber
Composition		E_s (GN/m^2)	Tenacity (MN/m^2)	E_s (GN/m^2)
AN/VBr	(4.2%)	18.1	550	—
AN/VBr	(4.2%)	19.6	596	200
AN/VBr	(4.2%)	19.6	768	—
AN/VBr	(4.2%)	23.0	736	262
AN/VBr	(4.2%)	23.0	733	—
AN/MAN	(5.0%)	14.5	505	200
AN/MAN	(5.0%)	18.1	662	200
AN/MMA	(4.2%)	13.6	417	145
AN/MMA	(4.2%)	15.5	496	165
AN/MMA	(4.2%)	18.1	587	207
AN/MMA	(4.4%)	18.1	550	214
AN/MMA	(4.2%)	21.2	790	234
AN/VBr (4.1%) AN/MMA (4.4%) } 1:1		18.1	551	214
AN/VBr (5%)/MA (3.5%)		18.1	598	228
AN/VBr	(1.9%)	21.2	643	213
AN/VBr	(4.2%)	19.6	723	—
PAN		19.6	664	—
AN/iBuMA		18.1	528	—
PAN		16.8	446	—
AN/VBr	(6.4%)	15.5	501	—
PAN AN/VBr (20.7) } 8:2		14.5	492	—
PAN		14.5	474	—
AN/MMA	(4.4%)	13.6	365	—
AN/VBr	(14.1%)	12.6	447	—
AN/VBr	(20.7%)	12.6	393	—
AN/VA	(7.4%)	13.7	391	—
AN/VA	(7.4%)	13.6	380	—
AN/VA	(7.4%)	13.6	358	—
AN/VA	(7.4%)	12.7	342	—
AN/VA	(7.4%)	11.9	321	—
AN/VA	(7.4%)	11.4	280	—
AN/VA	(7.4%)	9.3	219	—
AN/VA	(7.4%)	8.8	179	—

which is otherwise difficult to obtain, is required for the calculation of the "orientation parameter" α. A value of $E_{s,o} = (4.9 \pm 1.6)$ GN/m^2 results from Fig. 17, which is in good agreement with a value of 5.4 GN/m^2 ($= 5.6$ tex $= 50$ g/denier) estimated by Moseley [58] for unoriented Orlon, in a different way.

Using Eq. (22), the orientation parameter can be calculated. For the fibers represented in Fig. 17, α varies from 0.79 for the best composition, AN/VBr (4.2%), to 0.27 for the poorest, AN/VA (7.4%).

4.4.2 Molecular Weight of Copolymer

Most precursor copolymers have a specific viscosity in the range of 0.140–0.170, corresponding to a molecular weight range (weight average) of roughly $(80-100) \times 10^3$.

Table 12 shows the influence of the molecular weight on carbon fiber properties. In all cases, an improvement was observed with increasing molecular weight, although most pronounced in AN/VBr and in blends containing VBr. However, an upper limit is set by the solubility of the polymer. Moreover, the higher molecular weight polymer must be spun from a lower concentration to maintain the solution viscosity in a range suitable for spinning. Present experience indicates that a specific viscosity from 0.16 to 0.18 is probably most desirable.

Table 12. Influence of precursor molecular weight (specific viscosity, η_{sp}) on carbon fiber properties: comonomer concentration: 4.0–4.2%; "standard screening conditions"

Precursor			Carbon Fiber Properties		
Composition		η_{sp}	E_s (GN/m^2)	σ (MN/m^2)	ϱ (g/cm^3)
AN/VBr		0.143	213	1800	1.59
AN/VBr		0.172	213	2090	1.67
AN/VBr		0.180	234	2200	1.64
AN/VBr \} AN/MMA	1:1[*]	0.143 0.143	213	1430	1.66
AN/VBr \} AN/MMA	1:1[*]	0.175 0.163	213	2060	1.64
AN/MMA		0.140	206	1510	1.65
AN/MMA		0.175	213	1551	1.59
AN/VA		0.140	155	1070	1.57
AN/VA		0.161	165	1210	1.66

[*] Fibers from 1:1 copolymer blends

4.4.3 Shrinkage Force

If a fiber is stabilized at constant length, in a continuous furnace, and a strain gauge is applied to the fiber, at the exit of the furnace, an overall shrinkage force is generally observed. The measured force is the resultant of the effects of entropic shrinkage, elongation in the thermoplastic phase, and "chemical shrinkage" (cf. Sect. 3.3). Interestingly, the intensity of the overall shrinkage force depends very markedly on the type of comonomer present in the fiber. Table 13 shows that,

Table 13. Overall shrinkage force during stabilization; standard screening conditions

Precursor				Stabilization	Carbon Fiber	
Comonomer	Weight-%	Mol-%	Molar Volume	Shrinkage force (MN/m²)	E_s (GN/m²)	σ (MN/m²)
VBr	4.3	2.2	71.6	16.7	213	1800
MAN	5.0	3.4	83.9	11.4	200	1410
VA	3.9	2.4	92.4	8.6	165	1200
MA	3.9	2.4	90.3	7.6	186	1380
MMA	4.2	2.3	106.0	5.0	206	1510

at approximately comparable molar concentration, the overall shrinkage force decreases with increasing molar volume of the comonomer.

Neither entropic, nor chemical shrinkage are expected to be very sensitive to the small amount of comonomer present. More probably, the greater plasticization by the larger comonomer molecules enhances the elongation in the thermoplastic phase, thus reducing the apparent shrinkage.

From this point of view, improved carbon fiber properties are to be expected from the fibers with highest apparent shrinkage force, where the loss of attained molecular order by relaxation is the lowest. This may be an additional reason for the good performance of the AN/VBr precursor. In fact, the carbon fiber properties of this precursor are the best of those compared in Table 13. However, there is no clear trend relating properties to shrinkage force. As discussed in the preceding sections, the whole process of carbon fiber making is too complex to expect a simple relationship.

For certain application it may be of interest to use precursor fibers with low overall shrinkage force. Thus, if a non-woven fabric would be subjected to continuous stabilization, one would certainly wish to maintain the over-all shrinkage force low to minimize strain at the bonding points. AN/MMA might be an interesting candidate for such applications.

4.5 Correlation between Stabilized Fiber and Carbon Fiber Properties

Stabilization is the most critical step in the whole process of carbon fiber making, and it would certainly be desirable to have a method providing a rapid indication whether or not a given stabilization regime warrants good carbon fiber properties. The tenacity of fibers stabilized under the standard screening conditions, show in fact a certain relationship with the tensile strength of the carbon fiber (straight line in Fig. 18). However, data obtained from AN/VBr precursors under different conditions (black circles in Fig. 18; cf. Sect. 5.3), indicate clearly that such simple relationship does not exist. One of the possible reasons is that stabilized fiber tenacity may not be sensitive to incomplete stabilization (core), or to overoxidation. In both cases the damaging flaws will build up only during carbonization.

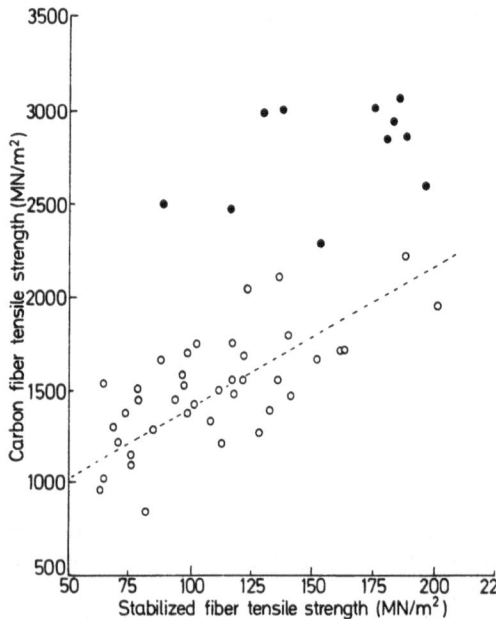

Fig. 18. "Correlation" between carbon fiber tensile strength and stabilized fiber tenacity. ● = AN/VBr optimized (cf. Sect. 5)

5 The Particular Case of the AN/VBr Precursor

5.1 Theory

The AN/VBr precursor evolved clearly as the top runner from the broad screening experiments reported in Section 4. A consideration of the general chemical and physical facts of carbon fiber formation — as discussed in Sections 2 and 3 — permits to discern a number of reasons for the special position of this precursor.

Vinyl bromide has two important properties which are not found combined in any of the other potential comonomers:
a) small molar volume
b) a very weak C-Br bond.
a) Due to the small volume of the bromine side chain, VBr as a comonomer does not essentially reduce the high molecular order, which can be obtained in stretched PAN fibers. Since good carbon fiber tensile properties are a direct consequence of molecular orientation (Sect. 3.5.2), the advantage is evident.

Intimately related to the greater molecular order is the relatively high melting point (cf. Fig. 4), which helps to prevent fiber fusion, permitting thus stabilization at higher temperatures than with other precursors. Since higher reaction temperature means higher reaction rate, there is a great potential for short stabilization times.

A third benefit from the small molecular volume of VBr is probably a reduced relaxation in the thermoplastic phase of stabilization (i.e. increased overall shrinkage force, cf. Sect. 4.4.3).

b) The weak C-Br bond (257 kJ/mol, cf. Table 2) is the first bond to break during heat treatment, at a temperature significantly below that required for main chain scission (main C-C bond in PAN: 297 kJ/mol).

The breaking of the C-Br bond gives radicals, without fragmentation of a macromolecule. The radicals may initiate the oligomerization of CN groups at a lower temperature, compared to the cases where main chain scission is the lowest energy radical source (PAN, AN/MMA, AN/MA, AN/VA etc.). This not only spreads the heat evolution over a greater temperature range, thus preventing fiber fusion (cf. Fig. 13), but also takes care of early crosslinking of macromolecules (cf. Formula 5 in Sect. 3.1). As a consequence, the formation of small fragments (fuel) is reduced in the temperature range of main chain scission.

The very reactive and mobile bromine radical most probably does not take part directly in the oligomerization reaction. Presumably it abstracts the nearest hydrogen it finds, creating a new carbon radical able to initiate the CN oligomerization:

$$R-CH_2-CH-R' + Br^* \rightarrow R-CH_2-C^*-R' + HBr$$
$$\quad\quad\quad\; | \quad\quad\quad\quad\quad\quad\quad\quad\quad | $$
$$\quad\quad\quad CN \quad\quad\quad\quad\quad\quad\quad\quad CN$$

Hydrobromic acid is very likely to initiate, additionally, the CN oligomerization by a different, ionic mechanism, see Formula 10 in Section 3.1.

The advantage of the small molar volume of VBr is shared by monomers like ethylene, vinyl chloride or vinyl iodide. But, the first two do not have a labile bond which could give radicals before main chain scission takes place (see Table 2). Vinyl iodide, on the other hand, has too labile a bond (C-I: 221 kJ/mol, Egger and Cocks [34]), which probably would not even survive spinning operations.

Vinyl chloride, in a polymer or copolymer, is known to decompose thermally by dehydrochlorination rather than by a radical break (cf. Formula 15). However, as mentioned in Section 3.1.1, HCl is considerably less prone to attack CN groups according to a reaction corresponding to Formula 10. Hence, VCl appears to be inferior to VBr for several reasons.

A weak bond, comparable to that in VBr, could in principle be introduced by other comonomers, such as styrene or α-methylstyrene (270 and 253 kJ/mol respectively, cf. Table 2). However, these weak bonds are located within the main chain. Moreover, these monomers have a large molar volume, hence a large negative effect on the molecular orientation. Thus, also from this point of view, VBr appears unique.

5.2 Optimization

Some work has been undertaken towards the optimization of the AN/VBr precursor. The work extends into all four relevant areas, viz. polymerization, spinning, stabilization and carbonization. Although probably far from a complete optimization, considerable improvement has been obtained with regard to the key factors — stabilization time, and carbon fiber tensile properties — as compared to the "screening conditions".

5.2.1 Polymerization

The effect of the molecular weight (specific viscosity) on the carbon fiber properties was discussed in Section 4.4.2, and the data available for AN/VBr precursors are included in Table 12. A range of η_{sp} of 0.17–0.18 appears to be the most recommendable.

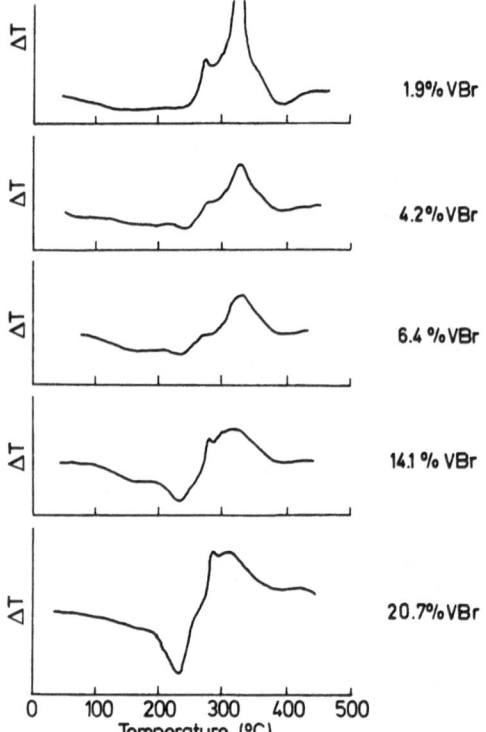

1.9% VBr

4.2% VBr

6.4 % VBr

14.1 % VBr

20.7% VBr   **Fig. 19.** Broadening of the DTA exothermic peak with increasing amount of VBr (weight %) in copolymers AN/VBr. (Heating rate 20 °C/min; air)

The VBr content in AN/VBr was also varied. Fig. 19 shows DTA data of AN/VBr precursors, with the VBr content varying from 1.9 to 20.7%. The increase of broadening and flattening of the exothermic peak, with increasing VBr content, is evident. (The endotherm preceding the exotherm is due to C-Br bond breaking and some dehydrobromination.) The suitability for carbon fiber formation, however, goes through a maximum. The lowest VBr copolymer fiber cannot be stabilized in short times (radical formation by C-Br bond breaking as well as radical multiplication by O_2, Formula 14, are too slow). The fiber with the highest VBr content tends to fuse. The carbon fiber properties are given in Table 14. When comparing the data, the following should be taken into account. Not all samples could be investigated under exactly the same conditions of spinning and stabilization. For AN/VBr (1.9%), the regime shown was the only one found suitable. The run with AN/VBr (4.2%) is one of the best made with this composition (cf. Table 18). The fiber from AN/VBr (6.4%) resulted from one of the only two experiments carried out with this precursor (the

Table 14. Influence of the VBr content in AN/VBr precursor fibers on the carbon fiber properties

Precursor			Stabilization			Carbon Fiber		
VBr (%)	η_{sp}	Spin stretch (X)	Time (min)	Profile[a]	Stretch (%)	E_s (GN/m^2)	σ (MN/m^2)	ϱ (g/cm^3)
1.9	0.150	13	45	A	0	213	1120	1.64
4.2	0.173	13	40	A	15	289	2950	1.73
6.4	0.139	10	30	A	7.5	234	2900	1.70
14.1	0.157	13	30	A	15	145	1610	1.58
20.7	0.111	6	several trials			no sample (fused)		

[a] See Fig. 14

other, with 5% stretch in stabilization, giving slightly poorer results). This composition has probably potential for further improvement, in particular since the molecular weight of the used copolymer was relatively low. The decline of properties for VBr > 14% is evident; the main reason is assumed to be reduced molecular orientation, although incomplete stabilization due to fusion problems cannot be completely ruled out. The optimum appers to be in the 4–6% (possibly 3–8%) range.

5.2.2 Spinning

Table 15 shows the effect of stretching during spinning, on the carbon fiber properties, in the particular case of an AN/VBr precursor with 4.2% of VBr. It can be seen that increased spinning stretch results in a small, but significant increase of the tensile strength σ of the carbon fiber, despite the fact that the higher stretched precursor fibers cannot take the same amount of stretch in stabilization.

Table 15. Influence of stretching during spinning, on carbon fiber properties. Precursor: AN/VBr (4.2%), $\eta_{sp} = 0.173$; stabilization: 50 min, profile B (cf. Fig. 16)

Spinning stretch (X)	Stabilization stretch (%)	Carbon Fiber Properties		
		E_s (GN/m^2)	σ (MN/m^2)	ϱ (g/cm^3)
6	30	260	2270	1.66
7	27	260	2320	1.66
9	15	290	2400	1.69
9	17	230	2460	1.68
13	10	260	2510	1.68

5.2.3 Stabilization

Stabilization has the variables time, temperature, and stretching. Table 16 summarizes four sets of experiments, for which stretching in stabilization was the only variable. In general, increased stretching brings increased tensile properties of the carbon fiber — up to a point where the fiber is damaged. The optimum depends on the degree

Table 16. Effect of stretching during stabilization on carbon fiber properties; AN/VBr precursors

VBr Content (mol%)	Spinning stretch (X)	Stabilization			Carbon Fiber Properties		
		Time (min)	Pro-file	Stretch (%)	E_s (GN/m²)	σ (MN/m²)	ϱ (g/cm³)
4.2	6	60	A	0	213	1800	1.59
				15	234	2070	1.59
6.4	10	30	A	5	213	2690	1.70
				7.5	234	2890	1.70
14.1	13	30	A	10	131	1390	1.54
				15	144	1500	1.58
				25	159	1800	1.57
4.2	13	30	A	5	262	2380	1.72
				10	262	2310	1.72

of stretching applied during spinning, as well as on the particular VBr content. Lower spinning stretch, and higher VBr content permit higher stabilization stretch. For the highly stretched AN/VBr (4.2%) precursor in Table 16, a stabilization stretch of 10% does not improve properties more than 5%. The question may arise whether it is more convenient to introduce the largest possible stretching in spinning or in stabilization. Table 17 tries to give an answer. Whereas an increase of the spin stretch by a factor of 2 leads only to an improvement of the tensile strength σ by 15%, a larger increase can be obtained from stabilization stretch. Nevertheless, a balanced application of both forms of stretching appears to be recommendable, since very high stabilization stretch tends to introduce problems with fiber fusion.

From Tables 14–17, the AN/VBr (4.2%) fiber, 13x stretched during spinning. emerges as the most promising carbon fiber precursor. It was also found that for this precursor a stabilization with 15% stretching, the temperature profile A (high entrance temperature, cf. Fig. 16), and a residence time of 40 min, leads to carbon fibers with very interesting tensile properties (see Table 17, last entry). For a further reduction of the residence time in the stabilization furnace, the temperature profile needs to be varied [72].

Table 18 shows that a mere reduction of the residence time (from 40 to 30 min), retaining profile A, brings about a slight deterioration of the carbon fiber tensile

Table 17. Spin stretch versus stabilization stretch. Precursor: AN/VBr (4.2%); $\eta_{sp} = 0.173$

Spinning stretch (X)	Stabilization			Carbon Fiber Properties		
	Time (min)	Pro-file	Stretch (X)	E_s (GN/m²)	σ (MN/m²)	ϱ (g/cm³)
6	60	A	0	213	2090	1.67
13	60	A	0	290	2300	1.70
6	40	A	25	290	2690	1.68
13	40	A	15	290	2950	1.73

Table 18. Influence of time/temperature regime during stabilization on carbon fiber properties. Precursor: AN/VBr (4.2%), $\eta_{sp} = 0.173$; spin stretch 13X; stabilization stretch 15%

Stabilization		Carbon Fiber Properties		
Time (min)	Temperature profile[a]	E_s (GN/m²)	σ (MN/m²)	ϱ (g/cm³)
40	A	290	2950	1.73
30	A	290	2600	1.72
30	D	290	3020	1.73
20	D	290	3020	1.73
15	D	290	2950	1.73
15	E	290	3000	1.72
11	E	260	2500	1.71

[a] See Table 19

Table 19. Temperature profiles in Table 18 (cf. Fig. 14)

Furnace zone	Temperature Profile (°C)		
	A	D	E
1	260	260	260
2	260 → 300	260 → 300	260 → 330
3	300	300 → 330	330

strength. Presumably, the stabilization is not quite accomplished under these conditions. A somewhat higher end temperature (profiles D and E, see Table 19) corrects for this defect, and carbon fibers with excellent tensile properties can be obtained with 30, 20 and 15 min stabilization time. An even shorter residence time (11 min) still results in interesting properties, though distinctly inferior to those of the 15–20 min samples.

5.2.4 Carbonization

Whereas all carbon fibers mentioned thus far have been carbonized under the "standard screening conditions" given in Section 4.2 (comprising a residence time of 6.7 min in a 75 cm long carbonization furnace, at 1400 °C), it was found that the

Table 20. Influence of carbonization rate on carbon fiber properties. Precursor fiber: An/VBr (4.2%), spinning stretch 13X; stabilization: 40 min, profile A, 15% stretching; carbonization: 1400 °C, furnace length 75 cm

Carbonization		Carbon Fiber Properties		
Rate (cm/min)	Residence time (min)	E_s (GN/m²)	σ (MN/m²)	ϱ (g/cm³)
11.2	6.7	290	2950	1.73
27.2	2.7	290	2940	1.71
45.5	1.6	290	2980	1.69

residence time could be considerably reduced without deteriorating the carbon fiber properties, when using the AN/VBr precursor. Table 20 shows data illustrating this point. The highest carbonization rate in the table corresponds to a residence time of less than 2 min.

6 Conclusions

Consideration of the chemistry and physics of the processes taking place during stabilization and carbonization indicated copolymers of AN and VBr as the most promising precursors for carbon fibers, in particular with regard to short, energy and cost saving stabilization times. The combined presence of two properties of VBr singled out this comonomer as uniquely suited for the purpose: 1) the low dissociation energy of the C-Br bond, permitting free radical formation at relatively low temperatures without disrupture of polymer molecules (early start of CN oligomerization, dissipation of the reaction exotherm), and 2) the low molar volume, limiting the melting point depression by the comonomer to acceptable values, hence permitting stabilization at a relatively high temperature from the very beginning (short stabilization time).

A broad screening of carbon fiber precursors, including binary copolymers of AN with a variety of comonomers, terpolymers and blends, in fact confirmed AN/VBr copolymers as the best precursor candidates.

Work towards optimization of the AN/VBr precursors and the process to transform them into good carbon fibers, led to the following specifications: The copolymer should have a VBr content of 4–6%, and a specific viscosity of 0.16 to 0.17. During spinning, the fiber should be drawn to ca 13 times its length. The stabilization can be carried out in a continuous furnace of 570 cm length, with a temperature profile ranging from 260 °C (entrance) to 310 °C (exit), a residence time of 15–20 min, and with 15% stretching. Carbonization of the stabilized fiber requires a residence time of ca 2 min in a continuous furnace at 1400 °C, at constant length.

From such procedure resulted carbon fibers with excellent tensile properties: tensile strength up to 3000 MN/m^2, and modulus up to 290 GN/m^2.

7 Experimental

7.1 Polymerization

The polymerizations were carried out in a pressurized continuous polymerization reactor of 8 l capacity (Autoclave Engineers). While most of the polymerization runs actually would not have required pressure, all runs were carried out at 2.5×10^5 Pa (N$_2$), which was convenient for the AN/VBr runs. The dispersion polymerization, at a water/monomer ratio of 3.5 (by weight), was catalyzed by a persulfate — bisulfite — iron system [73].

7.2 Spinning

The precursor fibers were wet-spun, from dimethylacetamide (DMAA) solutions, with 20–25% solids. Relatively low solubility of some of the copolymers required long holding time at elevated temperature (up to 112 °C) for dissolution, as well as heated storage and pumping operation. Since color formation was not critical for the present purpose, this did not represent any problem. Slow coagulation of the spun fiber in a bath of $DMAA/H_2O = 6/4$, at 40 °C, resulted in most cases in dense, void-free fiber structures at the spin-bath exit. The fibers were oriented using boiling water cascade stretch and/or steam stretch. The fibers contained 1,000 filaments, with a fiber weight (after drawing) of 0.14–0.15 tex/filament.

7.3 Stabilization

The continuous stabilization furnace was an insulated, 5.7 m long stainless steel tube of 2.5 cm diameter, which was capped at each end. The end caps had a 6 mm hole for fiber entrance and exit. The furnace was resistance heated. The temperature profile was established by wiring and controlling separately five heaters along the furnace. The bottom end of the vertically arranged furnace was chosen as the entrance end. A glass fiber pull-through line was dropped down through the furnace in order to string up the precursor fiber. Nip rolls and driving motors at entrance and exit guided the fiber through the furnace, and could be set for stretching if so desired. To successfully string up a fiber, the furnace had to be cooled down, at least to the intended entrance temperature; reheating took place at a rate of 0.5–1.0 °C per minute. These precautions were necessary to avoid burning of the fiber.

7.4 Carbonization

An inductively heated graphite tube furnace of 75 cm length and 1.5 cm inner diameter was used. Pyrolytic graphite discs at each end of the furnace minimized conductive heat loss. The tube was insulated with three layers of carbon felt, overwrapped with glass cloth. The furnace was continuously purged with nitrogen. Screening runs were carried out at 1400 °C, with a tension of ca 1 g/tex, and a residence time of 6.7 min. For work towards optimization of the AN/VBr precursor, maximum tension short of obtaining broken filaments was applied, to prevent loss of orientation during carbonization.

7.5 Carbon Fiber Characterization

The carbon fibers were characterized by their tensile strength σ, sonic modulus E_s and density ϱ.

The tensile strength σ was determined by measuring the strain-to-break, using Instron tensile testing equipment. Carbon fiber bundles (1000 filaments) were coated with a light film of epoxy resin to insure bundle integrity during testing. After curing, the bundles were cut to 15 cm length.

Due to the low breaking load of the bundles, Instron tensile moduli were not sufficiently reliable. Modulus data were therefore obtained by the sonic technique, using a modified PPM-5 (H. H. Morgan Company) dynamic modulus tester (Murayama [74]).

Density was determined by the floatation technique, using mixtures of tetrabromo-ethane ($\varrho = 2.96$ g/cm^3) and toluene ($\varrho = 0.87$ g/cm^3) such that a short length of fiber (0.5 cm) is just suspended in the vessel. The solution density is then measured with a gravitometer.

7.6 Glossary of Abbreviations

AAM	Acrylamide
AN	Acrylonitrile
i-BuMA	iso-Butylmethacrylate
DMAA	Dimethylacetamide
DTA	Differential thermal analysis
E_S	Sonic Modulus
EVE	Ethylvinylether
FTIR	Fourier Transform Infrared Spectroscopy
HEMA	Hydroxyethylmethacrylate
IA	Itaconic acid
k_B	Melting point depression constant
M	Monomer
MA	Methyl acrylate
MAN	Methacrylonitrile
α-MSty	α-Methylstyrene
MMA	Methyl methacrylate
NMAA	N-methylolacrylamide
PAN	Polyacrylonitrile
Sty	Styrene
T_g	Glass transition temperature
TGA	Thermogravimetric analysis
T_m	Melting point
VA	Vinyl acetate
VBr	Vinyl bromide
VCl	Vinyl chloride
VCl_2	Vinylidene chloride
η_{sp}	Specific viscosity
ϱ	Density
σ	Tensile strength

Acknowledgement. The authors acknowledge the excellent work of J. R. Sechrist (Polymerization), J. G. Brown (Spinning), K. H. Fulton (Stabilization) and J. G. Morrison (Carbonization), each of whom contributed essentially to the experimental part of this work. Thanks are also due to Dr. T. Murayama for the determination of the sonic moduli.

8 References

1. Edison, T. A.: US-Patents 223898 and 470925 (1879)
2. Hagen, H.: Glasfaserverstärkte Kunststoffe. Springer Verlag, Berlin, Heidelberg 1961
3. Frazer, A. H.: High Temperature Resistant Fibers. Interscience Publ., New York 1967
4. Shindo, A.: Studies on Graphite Fibers, Rep. Government Industr. Res. Inst. Osaka, Nr. 317 (1961)
5. Johnson, J. W., Phillips, L. N. and Watt, W.: Brit. Pat. 1, 110, 791 (1968)
6. Stuart, H. A.: Die Physik der Hochpolymeren, Vol. 1. Springer, Berlin, Heidelberg 1952
7. Henrici-Olivé, G. and Olivé, S.: Adv. Polymer Sci. *32*, 12 (1979)
8. Salame, M.: Polymer Preprints (ACS) *8*, No. 1, 137 (1967)
9. Hinrichsen, G.: Angew. Makromol. Chem. *20*, 121 (1974)
10. Flory, P. J.: Trans. Farad. Soc. *51*, 848 (1955)
11. Eby, R. K.: J. Appl. Phys. *34*, 2442 (1963)
12. Frushour, B. G.: Polymer Bull. *4*, 305 (1981)
13. Henrici-Olivé, G. and Olivé, S.: Coordination and Catalysis. Verlag Chemie, Weinheim. New York 1977
14. McCartney, I. R.: Mod. Plastics *30*, 118 (1953); Natl. Bur. Std. Circ. *525*, 123 (1953)
15. Grassie, N. and McNeill, I. C.: J. Polymer Sci. *27*, 207 (1958)
16. Grassie, N. and Hay, J. N.: J. Polymer Sci. *56*, 189 (1962)
17. Grassie, N. and McGuchan, R.: Europ. Polymer J. *7*, 1503 (1971)
18. Grassie, N. in Grassie, N. (ed.): Development in Polymer Degradation, Vol. 1, Chap. 5. Applied Science Publ., London 1977 (cf. earlier references therein).
19. Kausch, H. H.: Polymer Fracture, p. 85. Springer Verlag, Berlin, Heidelberg, New York 1978
20. Wöhrle, D.: Adv. Polymer Sci. *10*, 35 (1972)
21. Compagnon, P. L. and Miocque, M.: Ann. Chim. *5*, 11 (1970)
22. Johns, I. B.: Amer. Chem. Soc., Polymer Preprints *5*, 239 (1964)
23. Peebles, L. H. and Brandrup, J.: Makromol. Chem. *98*, 189 (1966)
24. Kargin, V. A., Kabanov, V. A., Zubov, V. P., and Zezin, A. P.: Dokl. Akad. Nauk. S.S.S.R. *139*, 605 (1962)
25. Kabanov, V. A., Zubov, V. P., Kovaleva, V. P. and Kargin, V. A.: J. Polymer Sci. C, *4*, 1009 (1964)
26. Howard, E. G.: U.S. Patent 2, 810, 726 (1957)
27. Osborn, J. H.: Ph. D. Thesis, University of Minnesota, Diss. Abstr. *19*, 2475 (1958)
28. Johnson, F., Panella, J. P., Carlson, A. A. and Hunneman, D. H.: J. Organ. Chem. *27*, 2473 (1962)
29. Urry, W. H., University of Chicago, private communication
30. Brandrup, J. and Peebles, L. H.: Macromolecules *1*, 64 (1968)
31. Brandrup, J. Kirby, J. R. and Peebles, L. H.: Macromolecules *1*, 59 (1968)
32. Liepins, R., Campbell, D. and Walker, C.: J. Polymer Sci. A-1, *6*, 3059 (1968)
33. Mullen, P. A. and Orloff, M. K.: Theoret. Chim. Acta *23*, 278 (1971)
34. Egger, K. W. and Cocks, A. T.: Helv. *56*, 1516 (1973)
35. Friedlander, H. N., Peebles, L. H., Brandrup, J. and Kirby, J. R.: Macromolecules *1*, 79 (1968)
36. Grassie, N. and McGuchan, R.: Europ. Polymer J. *7*, 1091 (1971)
37. Petcavich, R. J., Painter, P. C. and Coleman, M. M.: J. Polymer Sci., Polymer Phys. Ed. *17*, 165 (1979)
38. Grasselli, J. E., Wolfram, L. E. and Snavely, M. K. (1979) cited in Petcavich et al. (1979)
39. Balard, H. and Meybeck, J.: Europ. Polymer J. *14*, 225 (1978)

40. Menikheim, V. C.: U.S. Pat. 3, 817, 700 (1974)
41. Kubasova, N. A., Din, K. A., Geiderikh, M. A. and Shishkina, M. V.: Polymer Sci. USSR *13*, 184 (1971)
42. Geiderikh, M. A., Din, D. S., Davydov, B. E. and Karpacheva, G. P.: Polymer Sci. USSR, *15*, 1391 (1973)
43. Müller, D. J., Fitzer, E. and Fiedler, A. K.: Proc. Int. Carbon Fibers Conf., Plastics Inst. London, 1971, p. 10
44. Henrici-Olivé, G. and Olivé, S.: Polymer Bulletin *5*, 457 (1981)
45. Wentworth, G. and Sechrist, J. R.: J. Polymer Sci., Polymer Lett. *9*, 539 (1971)
46. Noh, I. and Yu, H.: J. Polymer Sci. Part B. *4*, 721 (1966)
47. Huron, J. L. and Meybeck, J.: Europ. Polymer J. *13*, 523, 553, 699 (1977)
48. Clarke, A. J. and Bailey, J. D.: Proc. Int. Carbon Fibers Conf., Plastics Inst., London 1974, p. 12
49. Peebles, L. H. in: H. Mark, N. Gaylord and N. Bikales (Eds.): Encyclopedia of Polymer Science and Technology. Suppl., Vol. 1, p. 2 Wiley, New York 1976
50. Coleman, M. M. and Petcavich, R. J.: J. Polymer Sci., Polymer Phys. Ed. *16*, 821 (1978)
51. Cook, A. G.: Enamines, Synthesis, Structure and Reactions. Dekker, New York 1969, Ch. 7
52. Watt, W., Johnson, D. J. and Parker, E.: Proc. Int. Carbon Fibers Conf., Plastics Inst., London, 1974, p. 3
53. Bell, J. W., and Mulchandani, R. K.: J.S.D.C. *81*, 16 (1965)
54. Hay, J. N.: J. Polymer Sci. A-1, *6*, 2127 (1968)
55. Warner, S. B.: J. Polymer Sci., Polymer Lett. *16*, 287 (1978)
56. Prescott, R. and Ott, I. W.: Germ. Pat. Appl. 1, 925, 489 (1969)
57. De Vries, H.: On the Elastic and Optical Properties of Cellulose Fibers. Schotanus and Jens, Utrecht 1953
58. Moseley, W. W.: J. Appl. Polymer Sci. *3*, 266 (1960)
59. Uchida, T. (1971) as cited by Bahl and Manocha [60]
60. Bahl, O. P. and Manocha, L. M.: Carbon *12*, 417 (1974)
61. Warner, S. B., Peebles, L. H. and Uhlmann, D. R.: Oxidative Stabilization of Acrylic Fibers III. Stabilization Dynamics. Industrial Liaison Program Publication, Massachusetts Institute of Technology 1978
62. Galin, M. and LeRoy M.: Europ. Polymer J. *12*, 25 (1976)
63. Bromley, J.: Proc. Int. Carbon Fibers Conf. Plastics Inst., London 1971, p. 23
64. Fiedler, A. K., Fitzer, E. and Rozploch, F.: Carbon *11*, 426 (1973)
65. Simitzis, J.: Colloid and Polymer Sci. *255*, 948, 1074 (1977)
66. Simitzis, J.: Ph. D Thesis, University of Karlsruhe, Germany 1975
67. Manocha, L. M., Bahl, O. P. and Jain, G. C.: Angew. Makromol. Chem. *67*, 11 (1978)
68. Bennet, S. C. and Johnson, D. L.: Carbon *17*, 25 (1979)
69. Crawford, D., cited by Bennett and Johnson [68]
70. Diefendorf, R. J. and Tokarsky, E. W.: The Relationships of Structure to Properties in Graphite Fibers, Technical Report AFML-TR-72-133, Part III, 1975
71. Marien, B. A.: J. Polymer Sci. Chem. Ed. *17*, 425, 435 (1979)
72. Henrici-Olivé, G. and Olivé, S.: Polymer Bulletin *6*, 229 (1982)
73. Peebles, L. H.: J. Appl. Polymer Sci. *17*, 113 (1973)
74. Murayama, T.: Dynamic Mechanical Analysis of Polymeric Material. Elsevier Scientific Publ. Comp., Amsterdam, Oxford, New York 1978
75. Stepanyan, A. Ye., Papulov, Yu. G., Krasnov, Ye. P. and Kurakov, G. A.: Vysokomol. Soyed. *A-14*, 2033 (1972)

Received November 10, 1982

On the Mechanism of Olefin Polymerization by Ziegler-Natta Catalysts

Vladimir A. Zakharov, Gennadii D. Bukatov and Yurii I. Yermakov
Institute of Catalysis, Siberian Branch of the USSR Academy of Sciences, Novosibirsk 630090, USSR

Novel data on the composition of active centers of Ziegler-Natta catalysts and on the mechanism of propagation and chain transfer reactions are reviewed. These data are derived from the following trends in the study of the mechanism of catalytic polymerization: a) determination of the number of active centers (mainly with the use of radioactive CO as a tag); b) analysis of the microstructure of polymers with the use of ^{13}C-NMR; c) analysis of specific features of highly active supported catalysts; d) quantum-chemical calculation of the electronic structure of active centers and their reactions.

1 Introduction

The two-component catalytic systems used for olefin polymerization (Ziegler-Natta catalysts) are combinations of a compound of a IV–VIII group transition metal (catalyst) and an organometallic compound of a I–III group non-transition element (cocatalyst) [1-4]. An active center (AC) of polymerization in these systems is a compound (at the surface in the case of solid catalysts) which contains a transition metal-alkyl bond into which monomer insertion occurs during the propagation reaction. In the case of two-component catalysts an AC is formed by alkylation of a transition metal compound with the cocatalyst, for example:

$$\text{Cl}_x\text{Ti} - \text{Cl} + \text{AlR}_3 \longrightarrow \text{Cl}_x\text{Ti} - \text{R} + \text{AlR}_2\text{Cl} \qquad (1)$$

$$\text{Catalyst} \qquad \text{Cocatalyst} \qquad \text{Active center}$$

: Surface of solid titanium chloride; Cl_x: chlorine ions of the first coordination sphere of titanium

There are also catalysts known in which ACs are formed during the direct interaction between the transition metal compound and the monomer without participation of special cocatalysts (so-called one-component catalysts). Examples of one-component catalysts are chromium oxide catalysts, lower halides of transition metals (titanium dichloride) and catalysts prepared by precipitating organometallic compounds of transition elements on oxide carriers [5].

For both types of catalysts polymerization proceeds by insertion of the olefin into an active transition metal alkyl bond:

$$\text{L}_x - \text{Mt} - \text{R} + n\ \text{H}_2\text{C} = \text{CHR}' \longrightarrow \text{L}_x - \text{Mt} - (\text{CH}_2\text{CHR}')_n - \text{R} \qquad (2)$$

$$\text{Mt: transition metal; } \text{L}_x\text{: ligands}$$

The following theoretical problems remain:

i) Elucidation of the AC composition. In the case of two-component catalysts a significant problem is to elucidate whether the cocatalyst effects the composition and the reactivity of the AC (alternative between monometallic or bimetallic active centers).

ii) Measurement of the number of AC_s and their localization on the catalyst surface. This problem is also related to the elucidation of the high activity of supported two-component catalysts as compared with traditional Ziegler-Natta systems.

iii) Determination of the AC reactivity (rate constants and activation energies for individual steps, especially for propagation and polymer chain transfer).

iv) Study of the mechanism of individual steps of polymerization: formation of the AC, polymer chain growth, and transfer. It implies the determination of elementary reactions composing individual steps. The study of the mechanism of the propagation reaction is necessary for the understanding of stereoregulation occurring in the polymerization of propylene and higher olefins.

First, the data on the overall polymerization kinetics were used to judge the above problems. The behaviour of catalytic systems was described by such parameters as activity, index of isotacticity, and average degree of polymerization. Later, data on the number of AC_s and their reactivity, the stereoregular structure of polymer chains, and on the molecular mass distribution of the polymer were also used to obtain a more detailed picture of the mechanism of catalytic polymerization. An increasing interest has recently been shown in quantum-chemical calculations of the electron structure of the AC and in a detailed characterization of elementary reactions.

In this article the problems of kinetics and the mechanism of olefin polymerization proceeding on solid Ziegler-Natta catalysts are discussed, using the novel data accumulated in polymer research:

1) *Number of $AC_s(C_P)$ and their reactivity in a propagation reaction (propagation rate constant k_P).* C_P data can be used to determine the mole fraction of active metal centers, the localization of active centers on the surface of the solid catalyst, and the role of individual components of the catalytic system in the formation of active centers. Systematic data on the influence of the catalyst composition and polymerization conditions on C_P and the rate constants of individual steps are important for the determination of the composition of the active centers and the elucidation of the mechanism of these steps. Various methods for determining C_P and k_P in catalytic polymerizations of olefins have been reported [6, 7]. A direct method for the determination of C_P is the radioactive tracer technique. In this method radioactive compounds react with the AC thus introducing radioactivity into the growing polymer chain. The use of radioactive alcohols is the classical example of this technique [8, 136]:

$$Mt - CH_2 - P + RO^3H \longrightarrow Mt - OR + {}^3H - CH_2 - P \qquad (3)$$

$$P: Polymer: chain$$

When using, however, two-component catalysts alcohols also react with inactive metal-polymer (aluminium-polymer) bonds which are formed in the chain transfer reactions with a cocatalyst. It is expedient to use the alcohol method only for catalytic systems and polymerization conditions for which the number of inactive metal-polymer bonds is low. Such a case is the polymerization of 4-methyl-1-pentene on vanadium trichloride activated with various organoaluminium compounds [9]. For this system the influence of catalyst composition and polymerization conditions on C_P and k_P was determined by quenching the polymerization with tritiated alcohol.

More universal is the method of C_P determination using selective tracers such as carbon monoxide and carbon dioxide which interact only with active metal = polymer bonds [10-13, 137]. Using tagged CO and CO_2 as quenching agents systematic data have been accumulated so far on the influence of the composition of catalysts (titanium chlorides with various organometallic compounds) and polymerization conditions on C_P and k_P values for the polymerization of ethylene and propylene [12, 14, 15].

2) *Comparative data of the mechanism of polymerization on active centers containing non-transition metals.* Olefin insertion into metal-carbon bond is also a key step not only for catalytic polymerization but also for olefin oligomerization with organoaluminium compounds. The latter process includes the same steps as the catalytic polymerization in the presence of transition metal compounds [16]:

$$\text{Initiation:} \quad \diagdown\!\!\text{Al}\!-\!\text{R} + \text{H}_2\text{C}\!=\!\text{CH}_2 \longrightarrow \diagdown\!\!\text{Al}\!-\!\text{CH}_2\text{CH}_2\!-\!\text{R} \qquad (4)$$

$$\diagdown\!\!\text{Al}\!-\!\text{H} + \text{H}_2\text{C}\!=\!\text{CH}_2 \longrightarrow \diagdown\!\!\text{Al}\!-\!\text{CH}_2\text{CH}_3 \qquad (5)$$

$$\text{Growth:} \quad \diagdown\!\!\text{Al}\!-\!(\text{CH}_2\text{CH}_2)_n\!-\!\text{R} + \text{H}_2\text{C}\!=\!\text{CH}_2 \longrightarrow \diagdown\!\!\text{Al}\!-\!(\text{CH}_2\text{CH}_2)_{n+1}\!-\!\text{R} \qquad (6)$$

$$\text{Termination:} \quad \diagdown\!\!\text{Al}\!-\!(\text{CH}_2\text{CH}_2)_n\!-\!\text{R} \longrightarrow \diagdown\!\!\text{Al}\!-\!\text{H} + \text{H}_2\text{C}\!=\!\text{CH}\!-\!(\text{CH}_2\text{CH}_2)_{n-1}\!-\!\text{R}$$

$$(7)$$

A comparative study of the insertion of olefins into transition metal- and non-transition metal-carbon bonds is essential for the clarification of the specific role of transition metals in catalytic reactions.

3) *Mechanism of stereoregulation on the basis of the data on polyolefin stereoregularity.* The structure of a polymer chain is the recording of events proceeding in the insertion of olefin molecules into an active metal-carbon bond. To understand the stereochemistry of the propagation reaction, the data on the stereoregular structure of polymer chains are important. Recently, for this purpose, ^{13}C-NMR spectroscopy has been extensively used [17].

4) *Studies of new catalytic systems.* Recently, several highly active Ziegler-type supported catalysts have been developed wherein the transition metal compound (usually titanium chloride or its complex with donor-type ligands) is supported on carriers [18, 19, 134, 135]. As the most effective supports, magnesium compounds, in particular anhydrous $MgCl_2$, are generally used. The activity of these catalysts is several orders of magnitude higher than that of the traditional systems.

5) *Quantum-chemical studies.* A quantum-chemical analysis provides a detailed picture of the nature of elementary reactions, proceeding with the participation of active centers, and of the electron structure of the latter. Pioneering work in this field by Cossee et al. [20, 21] is well recognized. In recent years, quantum-chemical studies were further developed [22-26], due to the possibility of using more advanced calculation methods (CNDO, MINDO, ab initio).

2 Number of Active Centers, their Composition and Reactivity

2.1 Measurements of the Number of Active Centers (C_p) and Propagation Rate Constants (k_p)

A proof of the quantitative determination of C_p using selective quenching agents is the conformity of the measured k_p values when i) different quenching agents were used and ii) the parameters of quenching were varied (concentration of agent, time of its contact with the polymerization medium, polymerization rate, polymer yield etc.).

^{14}CO and ^{14}CO$_2$ were proposed as selective quenching agents [10-13]. The insertion of carbon monoxide into transition metal-carbon σ-bonds proceeds via the step of coordination [27] (see also below). Carbon dioxide reacts with organometallic compounds according to the scheme:

$$L_x Mt - R + CO_2 \longrightarrow L_x Mt - O - \underset{\underset{O}{\parallel}}{C} - R \qquad (8)$$

Warzelhan and Burger [28] proposed to remove the cocatalyst before the introduction of $^{14}CO_2$ to avoid the possible influence of the cocatalyst on polymer radio-activity. However, two conditions have to be fulfilled to obtain correct data on C_P: i) the life-time of an active metal-polymer bond has to be sufficiently long; ii) the polymer must not contain Al—C bonds able to interact with a quenching agent. The possibility to use CO and CO_2 for the quantitative determination of C_P was illustrated in the case of one-component catalysts (titanium dichloride [29], supported organometallic catalysts [12]) when C_P values were independently determined by the use of other inhibitors including radioactive alcohol.

When carbon monoxide is introduced into the polymerization medium it is rapidly adsorbed at the active centers [13,30]. Then, CO is inserted into the metal-alkyl bonds which is manifested by the appearance of radioactivity in the polymer [13,31]:

$$L_x Mt - R \xrightarrow{^{14}CO} L_x Mt \overset{\overset{^{14}CO}{\downarrow}}{-} R \xrightarrow{^{14}CO} L_x Mt - \underset{\underset{O}{\parallel}}{^{14}C} - R \qquad (9)$$

Usually CO is used in excess with respect to the concentration of the AC_s, and adsorption of carbon monoxide on the AC_s leads to complete quenching of polymerization. If carbon monoxide is removed from the polymerization system, for example by vacuum, polymerization begins again [13,30]. However, polymer radioactivity (inserted ^{14}CO molecule) is not lost [13]:

$$L_x Mt \overset{\overset{^{14}CO}{\downarrow}}{-} \underset{\underset{O}{\parallel}}{^{14}C} - R \xrightarrow[(-^{14}CO)]{} L_x Mt - \underset{\underset{O}{\parallel}}{^{14}C} - R \xrightarrow{H_2C=CH_2} L_x Mt - CH_2 CH_2 - \underset{\underset{O}{\parallel}}{^{14}C} - R \qquad (10)$$

Such a mechanism of carbon monoxide interaction with active centers is compatible with the data on the slow copolymerization of CO with ethylene found for the ethylene polymerization by some one-component catalysts [32]. This copolymerization may proceed also in the case of two-component catalysts resulting in an increase of the number of radioactive tags in the polymer with time (see Fig. 1). Arguments have been given [13] that the rapid increase of polymer radioactivity in the initial period (5–10 min) is due to the insertion of the first ^{14}CO molecule into the active metal-carbon bond.

A further slow increase of the number of tags in the polymer can be due to:

a) copolymerization of ^{14}CO with the olefin and b) regeneration of the active center by an organoaluminium compound

$$Cl_x Ti - \underset{\underset{O}{\parallel}}{^{14}C} - P + AlEt_2 Cl \longrightarrow Cl_x Ti - Et + Et \overset{\overset{Cl}{\mid}}{Al} - \underset{\underset{O}{\parallel}}{^{14}C} - P \qquad (11)$$

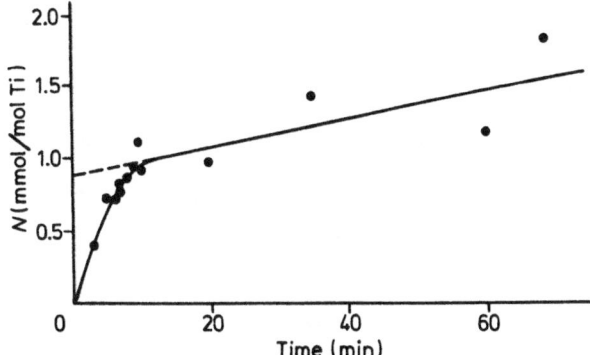

Fig. 1. Number of radioactive tags (N) in non-atactic polypropylene vs. the time of contact of ^{14}CO with the reaction medium after quenching of polymerization. Catalyst: δ-TiCl$_3 \times 0.3$ AlCl$_3$ + AlEt$_2$Cl; temp.: 70 °C; [titanium] = 1—2 mmol/l; [AlEt$_2$Cl] = 3 mmol/l, [Propylene] = 1 mol/l; ^{14}CO:Ti = 0.5–1.0 [13]

The more rapid increase of polymer radioactivity in the case of the system TiCl$_2$ + AlEt$_2$Cl + ^{14}CO compared with the system TiCl$_2$ + ^{14}CO may indicate the occurrence of a regeneration reaction [13].

Based on the number of radioactive tags in the polymer being at short (≈ 10 min) contact with ^{14}CO the following values of k_P for the polymerization of ethylene on titanium chloride catalysts were found [10,11,13,31,33–36] (80 °C, bulk and supported catalysts): $k_P \approx 1 \times 10^4$ l/(mol $\times$ s); for the propylene polymerization at 70 °C with bulk catalysts: $k_P \approx 1 \times 10^2$ l/(mol $\times$ s); with catalysts supported on MgCl$_2$: $k_P \approx 9 \times 10^2$ l/(mol $\times$ s). These values are compatible with several other data:

1) Similar k_P values ($\approx 0.5 \times 10^2$ l/(mol $\times$ s) for the propylene polymerization on titanium trichloride were obtained [37] using RO^{3}H as quenching agent and taking into account the contribution of inactive aluminium-polymer bonds (in particular at low polymer yield when the proportion of these bonds is small).

2) On the basis of the reported data on the achieved maximum activity of a titanium chloride catalyst supported on MgCl$_2$ ($\approx 5 \times 10^5$ g C$_2$H$_4$ per g Ti per h and per atm [19]) the minimum possible value of $k_P(k_P^{min})$ can be evaluated if one assumes that the number of AC$_S$ is equal to the total content of titanium in the catalyst. In this case, $k_P^{min} \approx 0.4 \times 10^4$ l/(mol $\times$ s). This value does not exceed and is close to the value of k_P found with a radioactive quenching agent used in the ethylene polymerization on a catalyst of this type.

3) Measurement of the molecular mass distribution of polyethylene prepared on a TiCl$_x$/MgCl$_2$ catalyst after a short time of polymerization (≈ 15 s) indicates the formation of a high molecular mass polymer; it proves the existence of AC$_S$ with a k_P of not less than 0.3×10^4 l/(mol $\times$ s) [38]. For the isotactic polymerization of propylene on supported MgCl$_2$ catalysts, similar data [34,36,39] show that the k_P value in this case has to be not less than 5.0×10^2 l/(mol $\times$ s).

4) The composition of ethylene-propylene copolymers allows the relative reactivities of olefins to be estimated. In the copolymerization of ethylene and propylene on titanium trichloride the reactivity ratio is ≈ 16 (at 75 °C) [40] should be higher in the homopolymerization of each component. This corresponds to the ratio (≈ 100 for

Table 1. Maximum polymerization rates (V^{max}), corresponding numbers of active centers (C_P^{max}), and k_P values for catalysts containing titanium chlorides [14]

Catalyst	Cocatalyst[a]	V^{max} in g polymer/ (mmol of Ti xh × atm)	$C_P^{max} \times 10^4$ in mol/mol of Ti	$k_P \times 10^{-2}$ in 1/(mol × s)
Polymerization of ethylene (75 °C)				
Bulk catalysts				
$TiCl_2$	—	8.0	1.1	120
$TiCl_n(Al)^{b)}$ (n < 2)	—	25.0	3.7	110
$TiCl_m(Mg)^{b)}$ (m ≦ 2)	—	150.0	19.0	130
$TiCl_2$	$AlEt_2Cl$	5.5	0.75	120
$\delta\text{-}TiCl_3 \times 0.3$ $AlCl_3$	$AlEt_3$	72.0	9.7	120
$\delta\text{-}TiCl_3 \times 0.3$ $AlCl_3$	$Al(i\text{-}Bu)_3$	96.0	11.2	140
Supported catalysts				
$TiCl_3$ on SiO_2	$AlEt_3$	960	110	140
$\alpha\text{-}TiCl_3 \times 0.5$ $AlCl_3$ on polyethylene	$AlEt_3$	720	89	130
$TiCl_4$ on $MgCl_xR_y$	$AlEt_3$	29000	3600	130
Polymerization of propylene^{c)} (70 °C)				
$TiCl_2$	—	0.005	0.013	0.76
$TiCl_2$	$AlEt_2Cl$	0.004	0.0085	0.94
$\alpha\text{-}TiCl_3$	$AlEt_2Cl$	0.11	0.33	0.71
$\delta\text{-}TiCl_3 \times 0.3$ $AlCl_3$	$AlEt_2Cl$	7.7	17	0.90
$\delta\text{-}TiCl_3 \times 0.3$ $AlCl_3$	$AlEt_3$	28	58	1.00
$\delta\text{-}TiCl_3 \times 0.3$ $AlCl_3$	$Al(i\text{-}Bu)_3$	36	80	0.90

a) Concentration: 3–10 mmol/l;

c) For propylene polymerizations data are given for polymer fractions insoluble in boiling ether (for $TiCl_3$) or n-heptane (for $TiCl_2$)

b) For these catalysts data were taken from Ref. [36]. Catalysts were prepared by reduction of $TiCl_4$ with $AlEt_3$ (Al:Ti = 2.5) at 25 °C or with $Mg(C_4H_9)_2$ (Mg:Ti = 1) at 0 °C;

bulk catalysts) of k_P values obtained by the method of selective quenching agents for ethylene and propylene polymerization (Table 1).

2.2 Data on the Number of Active Centers for Various Catalysts

The number of propagation centers significantly depends on the catalyst composition (Table 1). The wide variation of the activity (by 4 orders of magnitude) of titanium chloride-based catalysts is due to the change of number of AC_S. An exception is the propylene polymerization on $MgCl_2$-containing catalysts in which the change of activity is also due to a significant increase in the reactivity of active centers (Table 2).

The lowest C_P value is observed for the one-component $TiCl_2$ catalyst. The number of AC_S in one-component $TiCl_n(Al)$ and $TiCl_m(Mg)$ catalysts is much higher (see Table 1). This may be explained by the higher dispersity of these catalysts and also by the possible presence of Ti—C or Ti—H bonds due to the specific

Table 2. Number of active centers and propagation rate constants for isotactic (C_p^i and k_p^i) and atactic (C_p^a and k_p^a) polymerization of propylene[a]

Catalyst	$TiCl_m(Mg)$[b]	$TiCl_n(Al)$[b]	$TiCl_4/MgCl_2$-$AlEt_3$[b]	$TiCl_4 \cdot EB/MgCl_2$-$AlEt_3$[b]	Highly active supported catalyst[c]
V[d] in $g_{C_3H_6}/g_{Ti} \cdot h$ 500	40	27000	10000	35000
$C_p^i \times 10^3$ in mol/mol$_{Ti}$ 0.11	0.04	5.2	2.7	75
$C_p^a \times 10^3$ in mol/mol$_{Ti}$ 0.15	0.05	4.1	1.3	—
$k_p^i \times 10^{-2}$ in l/(mol×s) 4.1	1.0	7.4	8.7	5.0
$k_p^a \times 10^{-2}$ in l/(mol×s) 5.0	1.1	6.2	3.3	—

[a] Determined by using ^{14}CO as a radioactive quenching agent;
[b] Data taken from Refs. [34, 36]. Polymerization conditions: $T = 70$ °C, $[C_3H_6] = 1$ mol/l, $[AlEt_3] = 5$ mmol/l, [titanium] = 1 mmol/l and 0.05 mmol/l for bulk and supported catalysts, respectively;
[c] Data taken from Ref. [35]. Montedison-type catalyst; $T = 65$ °C; [propylene] = 0.29 mol/l, polymerization time: 5 min;
[d] Rate of propylene polymerization at the moment of the introduction of ^{14}CO into the reactor

technique of the catalyst preparation involving reduction of $TiCl_4$ with organometallic compounds.

Samples of α-$TiCl_3$, prepared by reduction of $TiCl_4$ with hydrogen, contain a low number of propagation centers. The C_P value of these well crystallized samples (the specific surface area according to BET is 3 m^2/g) is several per cent of the number of surface titanium ions. The low number of ACs is in agreement with the Cossee and Arlman concept [41,42] and the experimental data of Rodrigues et al. [43] on the localization of the AC$_s$ on the lateral faces and outlets of spiral dislocations on $TiCl_3$ crystals.

For δ-$TiCl_3 \times 0.3$ $AlCl_3$ samples, activated by dry grinding, the number of propagation centers is considerably higher and reaches 0.8×10^{-2} mol/mol of Ti (Table 1); this is close to the number of surface titanium ions (the value of the BET surface of catalysts 11 m^2/g). During the polymerization, however, the catalyst surface increases due to disintegration of the catalyst by the growing polymer [44–46]. This effect is the physical basis for the correlation between the activity and the size of primary crystallites established by means of the X-ray method [47,48]. The problem of the disintegration of the catalyst to primary crystallites or particles whose size is proportional to that of the primary crystallites has been analyzed [49]. Apparently, the "working" surface of the δ-$TiCl_3 \times 0.3$ $AlCl_3$ sample can attain 70–80 m^2/g after disintegration of the sample with an "initial" (BET) surface area of 11 m^2/g. In this catalyst the maximum number of propagation centers (0.8×10^{-2} mol/mol Ti; Table 1) corresponds to 10–15% of the total number of surface titanium ions. It should be noted that for δ-$TiCl_3$ samples, having a high degree of crystal structure disorder, the lateral faces can comprise a significant portion of the overall surface area. Taking into account these circumstances, data on the surface concentration of titanium ions taking part in the polymerization are not in contradiction to the concepts of the localization of propagation centers on lateral faces of $TiCl_3$ crystallites.

Catalysts prepared by supporting $TiCl_4$ on silica or polyethylene powder and activated by organoaluminium compounds before polymerization have an activity which is 10 times as high as that of bulk δ-$TiCl_3 \times 0.3$ $AlCl_3$ (Table 1). This increase is due to the increase of the number of AC_S. As it has been shown [50,51], the dispersed surface phase of $TiCl_3$ is an active component of these supported catalysts.

A drastic increase of activity and number of propagation centers (by more than hundred times in comparison with bulk $TiCl_3$) is achieved when non-solvated alkyl-magnesium chlorides or anhydrous $MgCl_2$ are used as supports. In these supported catalysts the number of active centers in ethylene polymerization achieves 36% of the total content of titanium (Table 1). Probably, the formation of a high concentration of active centers in these systems is due to titanium chloride stabilization on the support lattice owing to the similar parameters of the crystal lattice of titanium and magnesium halogenides [52]. To prepare highly active catalysts it is necessary to use supports with a high dispersity and a disordered crystalline structure. The increase of the order of crystalline structure of $MgCl_2$ samples at the same surface area leads to a decrease of the content of bound titanium chlorides [53]. It has been supposed [52] that titanium chloride interacts with surface defects of magnesium dichloride. The insertion of titanium chlorides into the surface layers of the support results in some increase of the structural order of its lattice which is confirmed by X-ray data [53,54]. Titanium chloride is strongly bound to the support surface and not removed by thermal treatment of the catalyst in vacuo at 200 °C [53,54].

The number of isotactic active centers (C_P^i) in propylene polymerization using one-component catalysts is close to the number of atactic active centers (C_P^a) (see Table 2). The k_P values for the two types of active centers are also similar (with the exception of the catalyst containing ethyl benzoate).

The maximum number of AC_S in propylene polymerization by supported titanium-magnesium catalysts may be estimated on the basis of the maximum activity of catalysts and k_P values ($\approx 10^3$ l/(mol $\times$ s). The initial activity of this catalytic system was as high [36] as 80 kg C_3H_6/(g Ti $\times$ h $\times$ atm) (in the presence of $Al(i\text{-}Bu)_3$ and ethyl p-methoxybenzoate), not less than 90% of isotactic polymer being formed. One can conclude that C_P^i is 0.07 mol/mol Ti. Taking into account C_P^a, the total number of active centers for this catalyst is $\approx 10\%$ of the content of titanium in the catalyst. These data are close to those obtained in ref. [35] (see Table 2).

Similar values of the number of active centers for C_2H_4 and C_3H_6 polymerization were found for identical samples of $MgCl_2$-containing catalysts. For example, the catalytic system $TiCl_4$/$MgCl_2$—$AlEt_3$ shows an activity of 70 kg C_2H_4/(g Ti $\times$ h $\times$ atm) in ethylene polymerization and of 50 kg C_3H_6/(g Ti $\times$ h $\times$ atm) in propylene poly-merization. Taking into account the monomer concentrations and k_P values ($\approx 10^4$/ (mol $\times$ s) for C_2H_4 and $\approx 10^3$ l/(mol $\times$ s) for C_3H_6), the number of AC_S may be estimated as 5.6×10^{-2} mol/mol Ti (for C_2H_4 polymerization) and 4.8×10^{-2} mol/mol Ti (for C_3H_6 polymerization).

The effect of polymerization conditions on C_P has been studied mainly for two-component systems based on titanium chloride [14,55,56] and vanadium chloride [9]. The number of propagation centers changes with the polymerization time proportion-ally to the reaction rate and is independent of the monomer concentration (0.2–2 mol/l at 70 °C). Most interesting is the effect of the polymerization temperature on C_P. It has been found that with rising temperature C_P increases (Table 3). In the

Table 3. Data on the overall activation energies (E_{ov}), activation energies of the propagation reaction (E_P), and the value ($E_c = E_{ov} - E_P$) determining the change of the number of active centers with temperature [55]. Catalyst: TiCl₃

Cocatalyst[a]	Monomer	E_{ov} kcal × mol^{-1}	E_P kcal × mol^{-1}	E_c kcal × mol^{-1}
AlEt₃	Ethylene	6.0	3.0	3.0
Al(i-Bu)₃	Ethylene	4.5	—	1.5
AlEt₂Cl		11.0	5.5	5.5
AlEt₃	Propylene	8.5	—	3.0
Al(i-Bu)₃		7.0	—	1.5

[a] Concentration of cocatalyst = 3×10^{-3} mol/l

general case the temperature dependence of C_P should be attributed to the change of the rates of the formation of propagation centers and of deactivation. Temporary deactivation of active centers may be caused by adsorption of an organoaluminium cocatalyst (OAC) [9,14,56]

$$\text{(12)}$$

The possibilities of these reactions are demonstrated by: i) the occurrence of chain transfer with the OAC which proceeds (see below) via the formation of A-type complexes (reaction (12)); ii) the reversible decrease of the polymerization rate with increasing OAC concentration [56]; iii) the decrease of C_P for the TiCl₂ samples upon the addition of OAC [10,12]; iiii) the effect of the type of the cocatalyst on the change of C_P with temperature (see Table 3).

For the catalytic system TiCl₃-AlEt₃ the equilibrium constant of reaction (12) has been determined to be equal to $3.9 \times 10^{-10} \exp(18500/RT)$ (in l/mol); the E_c value of this system can ranges from 0 to 10 kcal/mol, depending on the polymerization conditions (Fig. 2).

2.3 Reactivity and Stereoregulating Properties of Active Centers

2.3.1 Reactivity in the Propagation Reaction

In Table 1 the k_P values for various catalytic systems containing titanium chlorides are compiled.

Despite the significant differences in the catalyst activities, k_P is practically independent of the AlCl₃ content in TiCl₃ and TiCl₂, the type of organoaluminium

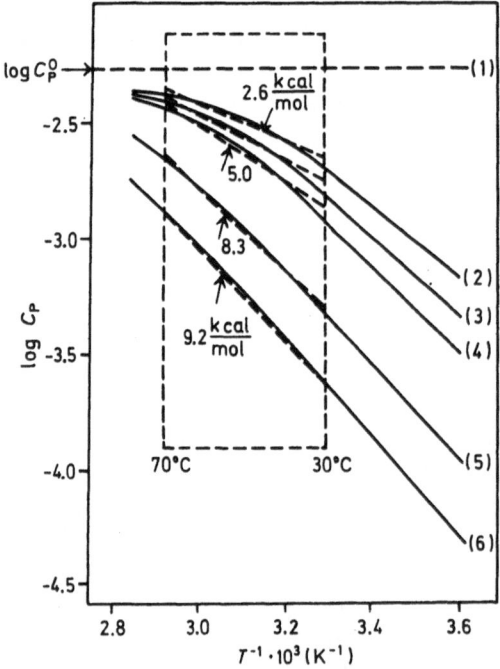

Fig. 2. Number of active centers vs. temperature at different cocatalyst concentrations. Concentration of AlEt$_3$ in mol/l. (1): 0; (2): 6×10^{-3}; (3): 1.5×10^{-2}; (4): 3×10^{-2}; (5): 3×10^{-1}; (6): 1.5. Data calculated [56] considering the reversible deactivation of active centers due to adsorption of AlEt$_3$. For the temperature region, given in the frame, average activation energies are indicated

cocatalyst, and of the type of the support. It is essential that k_P is the same for both the one- and two-component catalysts. For one-component (TiCl$_2$) catalyst the propagation centers are monometallic (containing no metal of cocatalyst). Hence, it may be supposed that the active centers of the two-component catalysts are also monometallic. This is consistent with the data on the independence of k_P on the cocatalyst nature for 1-butene polymerization on titanium and vanadium trichlorides [57].

For propylene polymerization on titanium chlorides containing MgCl$_2$ a significant increase of the reactivity of active centers in propagation has been found (Tables 1 and 2). Apparently, this increase in k_P reflects the influence of the magnesium ions on titanium ions in the second coordination sphere of titanium of the active center in MgCl$_2$-containing catalysts. Data obtained by use of the ^{14}CO quenching technique (Table 2) are in agreement with the evaluation of k_P from the molecular mass of isotactic polymer formed after a short time of polymerization. This evaluation [34, 36, 39] gives k_P^i of about 500–1000 l/(mol × s). The influence of the change of the composition of the second coordination sphere of titanium in active centers on the polymerization of ethylene (no change of k_P) and propylene (increase of k_P) is very intriguing and has not been clarified so far.

Similar reactivities of isotactic and atactic active centers have been found both for one-component (TiCl$_n$(Al) and TiCl$_m$(Mg)) and two-component (TiCl$_4$/

$MgCl_2-AlEt_3$) catalytic systems. Moreover, similar data [10] were obtained for $TiCl_2$ and $TiCl_2-AlEt_2Cl$ catalysts. The addition of ethyl benzoate to $MgCl_2$-supported catalyst does not change k_P^i but causes a decrease of k_P^a; thus, in this case, isotacticity depends not only on the ratio C_P^i/C_P^a but also on the ratio k_P^i/k_P^a.

2.3.2 Stereospecificity

Zambelli and Tosi [58] have extensively studied the stereochemistry of the propagation step in propylene polymerization on Ziegler-Natta catalysts. Specific features of this process are shown in Table 4. Cis-addition of the olefin to the active metal-carbon bond has been observed both in isospecific and syndiospecific polymerization. The olefin addition to the active bond proceeds with the participation of the primary ($L_xMt-CH_2-CHR-P$) and secondary ($L_xMt-CHR-CH_2-P$) carbon atoms of the growing polymer chain using isospecific and syndiospecific catalysts, respectively.

Table 4. Olefin polymerization according to ref. [58]

	Isospecific propagation	Syndiospecific propagation
Addition of olefin to the Mt-C bond	cis	cis
Orientation of monomer during insertion	Primary C atom to metal atom	Secondary C atom to metal atom
Chiral center determining stereocontrol	Metal ion of active center	Last monomeric unit of growing polymer chain

This conclusion is confirmed by the $^{13}C=NMR$ data on the regioregularity of polypropylene [59-62]. As it was shown for isospecific polymerization

i) there is no sequences of methylene carbons of the type ⌐_⌐ $\left(\;\rule{0pt}{1em}\rfloor = -CH_2-CH(CH_3)-,\quad --=-CH_2-CH_2-\right)$ in the ^{13}C-NMR spectrum of copolymers of C_3H_6 and $^{13}C_2H_4$;

ii) the number of the addition defects of the type ⌐_⌐⌐ in polypropylene (including atactic fraction) is considerably lower than the number of end groups;

iii) the isobutyl end groups are formed when C_3H_6 is inserted into Mt-$^{13}CH_3$ bonds [62]. These data indicate a high regioselectivity of isospecific propylene polymerization which is not determined by the type (C_2H_4 or C_3H_6) of the last monomeric unit of the growing polymer chain.

In syndiospecific polymerization using soluble vanadium catalysts, defects of the chemical addition are noticeable, their number being consistent with the Markoff first — order distribution [59-61,63]. In this case regioselectivity as well as stereospecificity are greatly influenced by the possibility of the secondary insertion of monomer [58]. The chirality of the last monomeric unit of the growing polymer determines the stereocontrol of syndiotactic polymerization. The introduction of an achiral (ethylene) unit results in the loss of stereocontrol [60,64]. As active center of syndiospecific polymerization the pentacoordinated vanadium-III ion was proposed [65]:

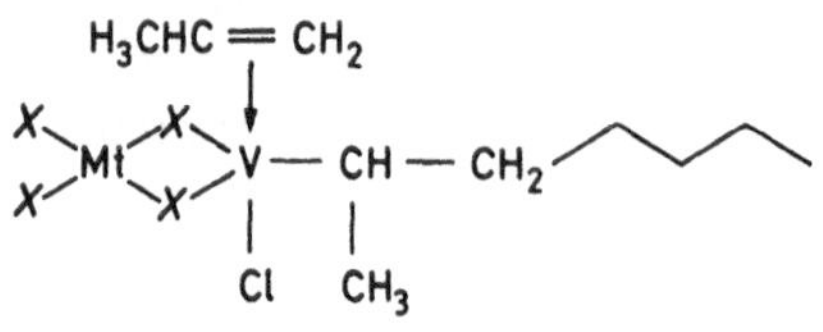

$$H_3CHC = CH_2$$

(Mt: Al, V; X: Cl, alkyl)

In isospecific polymerization the stereocontrol does not depend on the chirality of the last monomeric unit. It was found [60,64] that some ethylene units in ethylene-propylene copolymers ($TiCl_3$ as a catalyst) are surrounded by isotactic blocks of the same stereoregularity:

In this case the stereocontrol is due to the chirality of the active center. However, it was found [60] that isospecific addition of propylene occurs when the alkyl group attached to metal is more bulky than the methyl group (Table 5). Insertion of C_3H_6 into the Ti$-^{13}CH_3$ bond is not stereospecific and results in the formation of an equimolar mixture of threo- and erithro-isomers:

(12a)

(12b)

End groups of the type

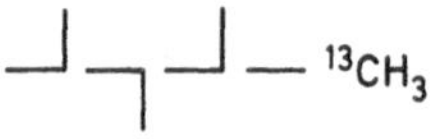

(12c)

(12d)

are practically absent. This confirms the high stereospecificity of propylene addition to the Ti$-$C$-$C$-^{13}$C bond in contrast to the addition to the Ti$-^{13}CH_3$ bond. Some stereocontrol ($\approx 70\%$ stereospecificity) was observed in the C_3H_6 insertion into the Ti$-^{13}CH_2CH_3$ bond. The same stereospecificty was obtained for the addition of propylene to the Ti$-^{13}CH_3$ bond using δ-TiCl$_3-$Al($^{13}CH_3$)$_2$I as catalyst. The stereospecific addition in the case of the Ti$-CH_2CH_3$ group was explained [66,68] by the influence of the β-carbon atom. The greater steric hindrance by C_β (in comparison with H of the Ti$-CH_3$ bond) results in the adsorption of olefin by the same face. The use of bulky alkyl groups (isobutyl, polymer chain), which cannot rotate freely about the Ti$-$C bond, results in a fixed position of C_β and a high stereospecificity of the active center.

Table 5. Stereospecificity of the olefin addition to titanium-alkyl bonds[a]

Catalytic system	Monomer	Active bond	Ratio threo:erythro[b]	Isospecificity of addition
δ—TiCl₃—Al(¹³CH₃)₂—Zn(¹³CH₃)₂[c]	C₃H₆	Ti—¹³CH₃	1:1	50
δ—TiCl₃—Al(¹³CH₃)₂I	C₄H₈	Ti—¹³CH₃	1:1	50
δ—TiCl₃—Al(¹³CH₂CH₃)₃—Zn(¹³CH₂CH₃)₂[c]	C₃H₆	Ti[d]—¹³CH₃	≈ 2:1	≈70
δ—TiCl₃—Al(¹³CH₃)₃—Zn(¹³CH₃)₂[c]	C₃H₆	Ti—¹³CH₂CH₃	≈ 2:1	≈70
	C₃H₆	Ti—CH₂CH(CH₃)¹³CH₃[e]	≧10:1	≧90

a) Determined by Zambelli et al.[66] according to ¹³C—NMR spectra of the end groups of the polymer; b) Threo: erythro ratio of ¹³C labelled end groups relative to the substituents of adjacent monomeric units; c) ZnR₂ was used as a transfer agent to increase the content of the end groups in the polymer; d) Apparently, a portion of the chlorine ligands of the first and/or second coordination sphere of the titanium ion is substituted by iodine[67]; e) The bond is formed after the addition of one C₃H₆ molecule to the Ti—¹³CH₃ bond

Table 6. Stereoisomer composition of polypropylene formed on one- and two-component polymerization catalysts containing titanium chlorides[70-72]

Catalyst	Polymer fraction solubility	content in wt-%	mmmm	mmmr	rmmr	mmrr	mmrm	rmrr	rmrm	rrrr	rrrm	mrrm	Ref.
δ—TiCl₃—AlEt₂Cl	Fraction soluble in boiling pentane	9	25	13	≈2	13	17	13	≈2	14	9	5	70
β—TiCl₃—AlEt₂Cl	Fraction soluble in boiling pentane	31	16	11	≈3	11	16	16	6	18	10	≈4	70
TiCl₂	Fraction soluble in boiling n-heptane[b]	18	49	10	≈2	10	6	11	6	10	6	5	71
TiCl₂—AlEt₃	Fraction soluble in boiling n-heptane[b]	15	57	11	≈2	11	8	9	≦1	9	4	5	71
δ—TiCl₃—AlEt₂Cl	Fraction soluble in boiling n-heptane[b]	1.5	55	7	—	7	13	7	—	11	3	3	70
β—TiCl₃—AlEt₂Cl	Fraction soluble in boiling n-heptane[b]	9	52	8	10	8	9	9	8	6	2	3	70
TMC[d]—AlEt₃	Fraction soluble in boiling n-heptane[b]	8.6	68	10	—	10	10	3	8	3	2	4	72
TiCl₂	Fraction insoluble in boiling n-heptane[c]	26	≈92	≈5	—	≈5	—	≈2	—	—	—	—	71
TiCl₂—AlEt₃	Fraction insoluble in boiling n-heptane[c]	54	92	4	—	4	4	—	—	3	—	—	71
δ—TiCl₃—AlEt₂Cl	Fraction insoluble in boiling n-heptane[c]	88.5	≧85	≦4	—	≦4	≦4	—	≦1	≦5	—	≦2	70
β—TiCl₃—AlEt₂Cl	Fraction insoluble in boiling n-heptane[c]	47.5	≈90	≈4	—	≈4	≈4	—	—	≈2	≈2	2	70
TMC[d]—AlEt₃	Fraction insoluble in boiling n-heptane[c]	47.8	89	4	—	4	4	1	—	—	—	2	72

a) Determined by ¹³C—NMR spectra of methyl groups; m: isotactic (meso) diad, r: syndiotactic (racemic) diad; b) Fraction insoluble in boiling ether (for TiCl₂) or n-hexane (for TiCl₃ and TMC); c) For TiCl₂ fraction was soluble in boiling n-octane; d) TMC — titanium-magnesium catalyst: TiCl₄/MgCl₂/C₆H₅COOC₂H₅

The effect of the last monomeric unit of the growing polymer chain on the stereospecificity of the olefin addition has been confirmed by the calculation of the energy of non-bonded interactions [69] and by quantum-chemical calculations (see section 5.2). Corradini et al. [69] have analyzed the possibility of the C_3H_6 π-complex formation on the octahedral titanium ions located on different faces of α- and γ-TiCl₃. The possibility of the coordination by both faces of the propylene molecule was studied. It was shown that active centers on the lateral faces of α-TiCl₃ and γ-TiCl₃ may be regioselective (primary insertion of propylene) ruther than stereospecific (no predominant C_3H_6 coordination by one face). In the case of active centers located on the edges of the layered modifications of TiCl₃, C_3H_6 is coordinated with the more accessible (outward) coordination sites of the titanium ions predominantly; the polymer chain is then located on the less accessible (inward) octahedral site. This position of the polymer chain results in a fixed orientation of the first carbon-carbon bond of the polymer chain due to its non-bonded interaction with the TiCl₃ surface. This may explain the predominant coordination of propylene molecules by one face and the stereospecificity of such type of active centers.

If the active centers of various catalysts differ in stereospecificity, one should expect differences in the stereoregular structure of the resulting polymer. For two-component isospecific catalysts Wolfsgruber et al. [70] have shown that the stereochemical composition of polypropylene fractions with similar solubility is almost independent of the type of transition metal chlorides (TiCl₃, VCl₃, ZrCl₄). The same stereochemical composition of polypropylene fractions was found for one- and two-component catalysts based on $TiCl_2$ [71] ($TiCl_2$ and $TiCl_2$-AlEt₃, respectively) as well as for the highly active supported catalyst TiCl₄/MgCl₂/C₆H₅COOC₂H₅/AlEt₃ [72] (Table 6). Using the last catalyst no change of stereoregularity of isotactic polypropylene was found [72] when AlEt₃ was replaced by AlEt₂Cl, MgCl₂ by MnCl₂, and ethyl benzoate by methyl acetate.

Thus, the experimental data show that the composition of catalytic systems does not influence the stereoregularity of the corresponding polymer fractions but only their relative content. Hence, the stereospecificity of the active centers of these catalysts is the same including the one-component catalyst $TiCl_2$. This confirms the concepts on the monometallic character of AC_S in the Ziegler-Natta catalysts. The possible existence of chiral titanium atoms on the titanium chloride surface was studied by Cossee and Arlman [41].

For the catalysts of different composition two polypropylene fractions can be singled out which differ greatly in their stereoregular structure (Table 6): a) a fraction insoluble in boiling n-heptane with an insignificant amount (1–2%) of steric defects of the type

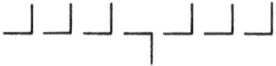

in the isotactic chain; b) a fraction soluble in boiling pentane (ether) with an approximately equal number of m and r diads exhibiting a complicated distribution when the formation of homosequences (mmm ... or rrr ...) is favored over that of heterosequences [70]. The formation of these fractions may be explained by the existence of two types of active centers on the catalyst surface: a) highly isospecific and b) non-stereospecific centers. The formation of an intermediate fraction (soluble (e.g. owing to the monomer or OAC adsorption near the nonstereospecific center) in boiling n-heptane) consisting of stereoblock macromolecules can be assumed to be due to the temporary conversion on non-stereospecific centers into isospecific ones which enhances steric hindrances, thus ensuring the one-mode olefin coordination to the active center.

Since for various catalytic systems only the relative content of different fractions changes (e.g. from 25 to 98.5% for a fraction insoluble in boiling n-heptane [70,71]) without changing their stereoregularity, the composition of catalytic systems influences the relative amount of isospecific and non-stereospecific centers. The reactivities of these centers (rate constants of propagation of isotactic and atactic polymers) for the titanium chloride-based catalysts are similar (Table 2 and Ref. [10]).

2.4 Role of the Organoaluminium Cocatalyst

The nature of the cocatalyst in the Ziegler-Natta systems exerts a significant effect on the polymerization rate and isotacticity of the polymer product [3,73–75]. Interesting data were obtained by Keii et al. [76,77] when the cocatalyst was changed during polymerization. The pronounced effect of the cocatalyst is the basis of the concept on the bimetallic composition of AC_S according to which the latter are represented, for example, as binuclear complexes including titanium and aluminium.

However, the comparative data on k_P (Table 1) and the stereoregularity of polymer fractions (Table 6) for one- and two-component catalysts based on titanium chlorides indicate that the cocatalyst does not influence the reactivity and stereospecificity of the propagation centers. Its effect on the overall polymerization rate is apparently due to the change in the total number of active centers and the ratio of isospecific and non-stereospecific centers.

The effect of the cocatalyst is reflected in various steps of polymerization. In the formation of propagation centers the nature of the cocatalyst can greatly influence the total number of AC_S and the relative number of stereospecific and non-stereospecific centers. In the chain growth process temporary deactivation of AC_S may occur due to the adsorption of the cocatalyst reaction (12)). Thus, the addition of $AlEt_2Cl$ to $TiCl_2$ markedly decreases the number of active centers in atactic propylene polymerization with a slight change in the number of isospecific active centers; the isotacticity of propylene increases from 25 to 55% [10]. This mechanism of cocatalyst action has been confirmed by experiments, involving the addition of triphenyl-phosphine instead of organoaluminium compounds to $TiCl_2$; the decrease in activity is also accompanied by an increase of polypropylene isotacticity [78].

In chain termination reactions, the cocatalyst acts as a transfer-agent (see below). It can also re-initiate the propagation centers subsequent to other chain termination processes: i) after deactivation of the centers by the impurities present in the polymerization medium; ii) after monomer insertion, binding a secondary carbon atom to the metal ion instead of the normal primary insertion:

$$\text{Mt} - \text{C} - \text{C} - (\text{C} - \text{C})_n \diagup\diagdown\diagup + \text{AlR}_3$$

Poorly active
center

$$\text{Mt} - \text{R} + \text{R}_2\text{Al} - \text{C} - \text{C}\diagup\diagdown\diagup \qquad (13)$$

Highly active
center

3 Mechanism of the Propagation Reaction

The propagation reaction is the main step of catalytic polymerization. The study of its mechanism implies the elucidation of its elementary steps including the determination of the rate-determining step. The most important properties of AC_s necessary for the reaction to proceed include: i) the presence of a metal-carbon bond; ii) coordinative unsaturation of a metal ion in the AC. Here, we discuss the data on the mechanism of the propagation reaction with respects to the AC containing the transition metal-carbon bond and also for organoaluminium compounds, i.e. when the AC contains a non-transition metal carbon bond.

3.1 Insertion of Olefin into the Transition Metal-Carbon Bonds

3.1.1 Elementary Steps of the Propagation Reaction

In Natta's early publications [79] a two-step mechanism of propagation has been suggested which includes olefin coordination on a transition metal ion and insertion of the coordinated monomer into an active metal-carbon σ-bond:

$$\text{Ti} - \text{C}\diagup\diagdown\diagup + \overset{}{\underset{}{\text{C}}} = \overset{}{\underset{}{\text{C}}} \underset{k_{-1}}{\overset{k_1}{\rightleftharpoons}} \text{Ti} \begin{array}{c} \text{C} = \text{C} \\ | \\ \text{C}\diagup\diagdown\diagup \end{array}$$

$$\downarrow k_2$$

$$\text{Ti} - \text{C} - \text{C} - \text{C}\diagup\diagdown\diagup \qquad (14)$$

The experimental k_P value of the two-step mechanism is described by

$$k_P = \frac{V}{C_P[M]} = \frac{k_1 \cdot k_2}{k_2 + k_{-1} + k_1[M]} \tag{I}$$

Depending on the ratio between rate constants of elementary steps (k_1, k_{-1}, k_2) and the monomer concentration, the following cases can be distinguished:

$$k_2 \gg k_{-1} + k_1[M]; \qquad k_P = k_1 \tag{II}$$
$$k_{-1} \gg k_2 + k_1[M]; \qquad k_P = k_2 k_1/k_{-1} \tag{III}$$
$$k_1[M] \gg k_2 + k_{-1}; \qquad k_P = k_2/[M] \tag{IV}$$

The normally observed first-order kinetics of polymerization with respect to monomer concentration corresponds to the cases II or III.

In Boor's review [3] a one-step propagation mechanism was discussed the driving force of the reaction being assumed to be the polarization of the Ti—C and C=C bonds:

Transition state

According to this mechanism a first-order kinetics of the propagation reaction with respect to monomer should also be observed.

There are several experimental data in the favour of the two-step mechanism of propagation. Natta [79], in his studies of the relative reactivity of various styrene derivatives, has found that the rate of propagation is governed only by the nature of a monomer entering into a polymer chain and is independent of the nature of the last unit of the chain end bound to the titanium ion. On the basis of these data Natta suggested that the propagation reaction proceeds via two steps — coordination and insertion, the former being the rate-determining one. For several Ziegler-Natta catalysts (at low temperature of polymerization and high monomer concentration) a transition from first to zero-order kinetics of the propagation reaction with respect to monomer was observed [80-82]. This corresponds to case (IV) and does not apply to a one step mechanism. For supported organometallic catalysts containing Ti(III) and Zr(III) ions as active centers, ESR studies show that the propagation centers reversibly adsorb ethylene at low temperatures [83,84,91]. These data may be considered as a direct evidence of monomer coordination on a transition metal ion of the active center.

A four-step mechanism for olefin polymerization with the participation of a carbene and metalcyclobutane complexes has been proposed to describe both α-olefin polymerization on Ziegler-Natta catalysts and olefin metathesis [85]:

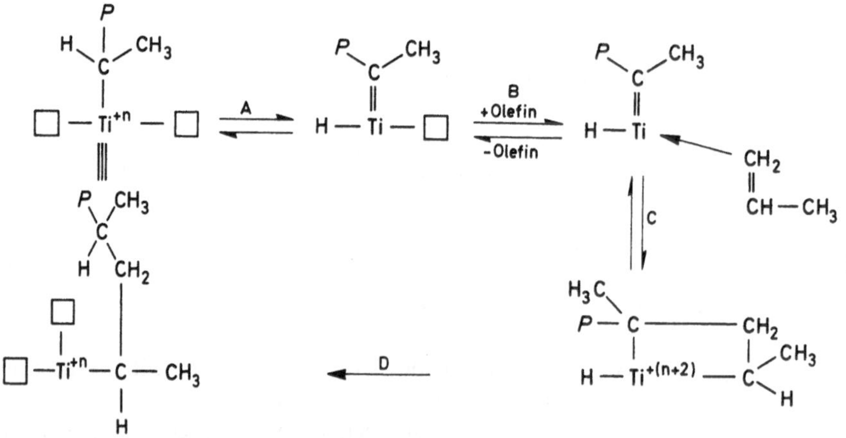

$$
\begin{array}{c}
\text{(16)}
\end{array}
$$

According to this mechanism migration of an α-hydrogen atom occurs via intermediate formation of a Ti-H bond; if step D is slow, olefin metathesis occurs:

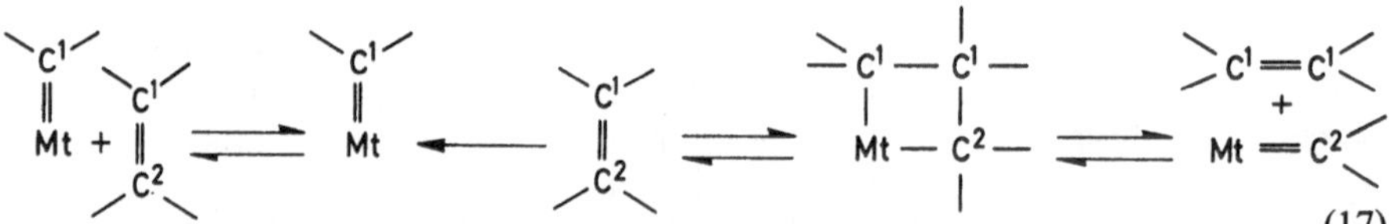

$$
\begin{array}{c}
\text{(17)}
\end{array}
$$

Many experimental data are known [86] confirming that olefin metathesis (including cycloolefin polymerization) proceeds via carbene intermediates.

However, this concept cannot be transferred to the polymerization of l-alkenes. Recently, critical remarks on the carbene mechanism of ethylene insertion into the $Co-CH_3$ bond were published [87]. Zambelli et al. [62,88] performed an elegant study on the discrimination of the carbene mechanism for the stereospecific polymerization of propylene. According to the carbene mechanism the insertion of the first C_3H_6 molecule into the $Mt-^{13}CH_3$ bond have to result in the formation of chain with an isobutyl end group: i) enriched by ^{13}C in the methyl and methylene groups (scheme (17a) [62]); ii) enriched by ^{13}C in the methyl group with threo- or erithro-configuration (see structures (12a) and (12b)).

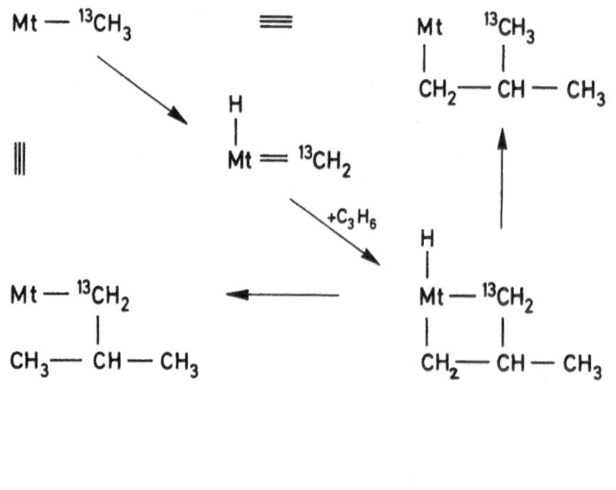

$$
\begin{array}{c}
\text{(17a)}
\end{array}
$$

For polypropylene prepared with the catalytic system δ-TiCl$_3$—Al(^{13}CH$_3$)$_2$I only end groups enriched by ^{13}C in the CH$_3$ fragments with prevailing threo-configuration were found [62]. These data are not compatible with the carbene mechanism.

Casey [89] studied the carbene mechanism for the case of a chiral environment of the transition metal in the active center. A chiral environment is possible for the catalytic system TiCl$_3$—Al(^{13}CH$_3$)$_2$I in comparison to the TiCl$_3$—Al(^{13}CH$_3$)$_3$ system (Table 5). This accounts for the prevailing threo-configuration of the ^{13}C-enriched methyl groups if the TiCl$_3$—Al(^{13}CH$_3$)$_2$ I system follows the carbene mechanism. However, it is not yet possible to explain the absence of ^{13}C in methylene fragments. Thus, the experimental data of both isospecife [62] and syndiospecific [88] polymerization of propylene are not consistent with the carbene mechanism. A non-contradicting explanation of the available experimental results is based on the mechanism of polymerization involving monomer insertion into the σ-metal-carbon bond.

Recently, McKinney [90] proposed the formation of a metal-cyclopentane intermediate for the addition of olefin due to the simultaneous coordination of two monomer molecules:

$$(17\,b)$$

According to quantum-chemical calculations [90] the formation of the 2,4-dimethyl-cyclopentane intermediate is favored. However, Zambelli et al. [68] reported that the formation of a metal-cyclopentane intermediate cannot explain the isotactic and syndiotactic polymerization on the base of the same type of mechanism. With some assumptions (for example, about the chirality of metal environment) the metal-cyclopentane mechanism can explain the formation of isotactic propylene. However, the driving force of the formation of the syndiotactic polymer is not evident at all.

3.1.2 The Rate-Determining Step of the Propagation Reaction

With a two-step mechanism and a first-order propagation reaction with respect to monomer, the two cases (II) and (III) are possible, propagation rate being determined either by the rate of monomer coordination ($k_P = k_1$) or by the rate of coordinated olefin insertion and its surface concentration ($k_P = k_2 k_1/k_{-1}$) (Table 7).

According to Natta [79], the rate-determining step of the polymerization of styrene and its derivatives is the coordination ($k_P = k_1$) since the nature of the last unit of the

Table 7. Activation energies (E_p) and frequency factors (A_p) corresponding to different cases of the first-order kinetics of the two-step propagation reaction

	Meaning of E_p	Meaning of A_p
Case II: $k_p = k_1$	E_1	A_1 $(0.3 - 3.0) \times 10^5$ 1/(mol × s)[c]
Case III: $k_p = k_2 \cdot k_1/k_{-1}$	$E_2 - q$[b]	$A_1 A_2/A_{-1} \ll A_1$[d]
Experimental data[a]	3–6 kcal/mol	(3 to 8) × 10^5 1/(mol × s)

[a] Experimental data on the olefin polymerization proceeding on titanium or vanadium chlorides [55,57]; data on A_p were calculated for k_p in the range of 10^2 to 10^4 1/(mol × s) [10,29,55];
[b] $q = (E_{-1} - E_1)$: heat of the coordination of olefin on the transition metal ion;
[c] Estimated according to the method of an activated complex [15];
[d] This expression follows from the supposition that $A_2 \ll A_{-1}$

growing chain end does not influence the rate of propagation. In earlier contributions to olefin polymerization the overall activation energy of 10–15 kcal/mol was considered to be the activation energy of the propagation reaction. This value was regarded as being too high if it corresponded to the monomer coordination as the rate-determining step. However, it was shown [55,57] that the real activation energy of the propagation reaction is lower (3–6 kcal/mol).

A comparison of experimental and calculated frequency factors (Table 7) of the propagation rate constant permits to draw some conclusions on the mechanism of propagation. For case (II) the explicit condition $A_p = A_1$ is expected. For case (III) one can expect that $A_p = A_1 \cdot A_2/A_{-1} \ll A_1$ since $A_2 \ll A_{-1}$. The relation $A_2 \ll A_{-1}$ accounts for the possible interrelation between the frequency factors of two monomolecular reactions: the formation of a complex four-centere intermediate during the insertion reaction compared with simple olefin desorption [55]. The better agreement between the calculated A_1 and experimental A_p values corresponds to the case for which the rate-determining step involves olefin coordination.

3.2 Insertion of Olefins into the Aluminium-Carbon Bond

3.2.1 Active Centers

Numerous kinetic studies [92−98] of the addition of olefins to the aluminium-carbon bond in trialkylaluminiums have provided conclusive evidence that the AC of this reaction is a monomeric form of the trialkylaluminium molecule.

The kinetics of the addition of olefin to the aluminium-alkyl bond is described by the equation

$$\frac{-d[R'{-}CH{=}CH_2]}{d\tau} = k_p[Al_2R_6]^{0.5}[R'{-}CH{=}CH_2] \qquad (V)$$

The necessity of a vacant coordination site for the addition of an olefin to AlR_3 is confirmed by the data on the effect of donor-type compounds (e.g. ether) on the reaction kinetics [99]. It has been found that in olefin addition only a monomeric form of triethylaluminium is active whereas the complex $Et_3Al \cdot Ph_2O$ is inactive. It has been shown that the presence of ether does practically not change the rate constant but drastically decreases the concentration of the monomeric form $AlEt_3$, due to the formation of an inactive $Et_3Al \cdot Ph_2O$ complex.

3.2.2 Mechanism of the Propagation Reaction

The participation of a monomeric form of the organoaluminium compound in the olefin addition may result in a preliminary interaction between the olefin and vacant p-orbitals of the aluminium ion with the formation of a π-complex before insertion of the olefin into the Al—C bond. The first-order kinetics of this process with respect to monomer implies that the concentration of the π-complexes with respect to the mono- mer OAC form is low; at elevated reaction temperatures these complexes were not identified by physical methods.

However, in several special cases data on the possible formation of π-complexes with a monomeric form of OAC have been obtained. Thus, NMR studies [100] have shown the formation of a π-complex by interaction of phenylacetylene with triethyl- aluminium. The heat of this reaction is 0.7 ± 0.2 kcal/mol. Taking into account the heat of dissociation of an Al_2Et_6 molecule (17 kcal/mol [101]) the strength of the π-bond between phenylacetylene and aluminium ion is about 9 kcal/mol. The rate of phenylacetylene insertion into the aluminium-ethyl bond is first order with respect to the concentration of the π-complex, according to reaction (18):

$$(18)$$

IR and NMR spectroscopy studies have shown [102-104] the formation of intra- molecular π-complexes:

Thus, data on the participation of the coordinatively unsaturated forms of OAC in the reaction with olefins and the data on the π-complex formation with double and triple bonds permit to conclude that the olefin insertion into the Al—C bond proceeds by a two-step mechanism.

3.2.3 Rate Constants of Olefin Addition to Al—C and Al—H Bonds

The activation energies of the addition of olefins to $Al(C_2H_5)_3$ increase from 12–15 to 16–19 kcal/mol in the series from propylene to 1-octene [94,95,97,98,105,106]. The

Table 8. Activation energies (E) of the addition of olefins to different organoaluminium compounds

Olefin	Organoaluminium compound	E kcal $\times$ mol^{-1}	Ref.
1-Octene	$\begin{cases} Al(C_2H_5)_3 \\ Al(CH_3)_3 \end{cases}$	18.4 20.5	96) 96)
Ethylene	$\begin{cases} Al(C_2H_5)_3 \\ Al(CH_3)_3 \end{cases}$	16.8 22.5	109,110) 109,110)

frequency factors (A) also increase in the some order approximately by an order of magnitude (log $A = 3.3$–5.0) [97].

Olefin addition to $Al(CH_3)_3$ proceeds more slowly than in the case of $Al(C_2H_5)_3$. This is due to the higher dissociation heat of $[Al(CH_3)_3]_2$ [107,108] and to the higher activation energy of the addition reaction (Table 8).

The interaction of ethylene with triethylaluminium has been studied in the gas phase under conditions under which the formation of 1-butene is predominant [109,110]. The resulting kinetic parameters of this reaction are given in Table 8. It should be taken into account that in this case for the calculation of the rate constants of the addition reaction several additional assumptions are introduced concerning the concentrations of various OAC forms and the ratio of the rate constants of various reactions. Apparently not all the assumptions are fulfilled and the activation energies of the ethylene addition listed in Table 8 are overestimated. Ziegler et al. [16,92] have found the following order of reactivity of olefin addition to organoaluminium compounds:

$$H_2C=CH_2 > H_2C=CHR > H_2C=CR_2 .$$

When passing from metal-carbon to metal-hydrogen bonds, the reactivity of olefin addition sharply increases due to the decrease in the activation energy (Table 9). The activation energy of the addition to the Al—H bond is practically the same for various olefins, and for the same olefin it increases in the order $R_2AlH < R_2GaH < R_2BH$.

Table 9. Activation energies of the addition of olefins to different metal-hydrogen bonds

Olefin	Organometallic compound	E kcal $\times$ mol^{-1}	Ref.
Ethylene 1-Butene 2-Methylpropene $\Big\}$	$(i\text{-}C_7H_9)_2AlH$	5.0–6.0	111,112)
2-Methylpropene	$\begin{cases} (i\text{-}C_7H_9)_2GaH \\ (i\text{-}C_7H_9)_2BH \end{cases}$	8.8 12.0	113) 114)

3.3 Comparison of Propagation Steps on Various Active Centers

The kinetic parameters of the propagation of olefin polymerization on different active centers are compiled in Table 10. Apparently, for both the transition and non-transition metal compounds the insertion of the olefin into Mt—C bonds proceeds with the participation of coordinatively unsaturated metalalkyl compounds via intermediate π-complexes. The higher reactivity of transition metal compounds compared with organoaluminium compounds is primarily due to the lower activation energy of the propagation step when Mt is a transition metal. Many facts indicate that polarization of the Mt—C bond does not determine the reactivity of metalalkyl compounds in olefin addition, e.g. due to the decrease of reactivity in the order

$$H_2C = CH_2 > H_2C = CHR > H_2C = C\overset{R_1}{\underset{R_2}{\diagdown}} \quad [92],$$ the increase of activation energies

when passing from $Al(C_2H_5)_3$ to $Al(CH_3)_3$ (Table 8), and the decrease of activation energies when passing from Al—C to Al—H bonds (Tables 8 and 9).

The activation energies of the addition of olefins to Mt—L bonds (where L is hydrogen or alkyl) do not correlate with the energies of these bonds (see Table 13).

Apparently, the reactivity of organometallic compounds in the addition of olefins to Mt—C bonds is determined by the capability of these compounds to coordinate olefins. The formation of intermediate π-complexes ensures further insertion of olefin by a concerted mechanism with a low activation energy. Thus, a high reactivity of active centers, containing a transition metal, comparable to the reactivity of the radical active centers, is achieved. The activation energy of the propagation in olefin polymerization on catalysts containing transition metals (2–6 kcal/mol) does not exceed its value for the radical polymerization (Table 10).

Table 10. Kinetic parameters for the propagation reaction of olefin polymerization on various active centers

Catalyst	Fragment of AC	Monomer	Propagation rate constant in $l/(mol \times s)$	Ref.
CrO_3/SiO_2	Cr—C	C_2H_4	$4.5 \times 10^5 \exp(-4500)/RT$	[5]
$TiCl_3 + AlEt_3$	Ti—C	C_2H_4	$8.0 \times 10^5 \exp(-3000)/RT$	[55]
$TiCl_3 + AlEt_2Cl$	Ti—C	C_3H_6	$3.0 \times 10^5 \exp(-5500)/RT$	[55]
$TiCl_3 + AlEt_2Cl$	Ti—C	C_4H_8	$A \cdot \exp(-4500)/RT$	[57]
$VCl_3 + AlEt_3$	V—C	C_4H_8	$A' \cdot \exp(-2300)/RT$	[57]
$AlEt_3$	Al—C	C_3H_6	$2 \times 10^3 \exp(-12000)/RT$	[97]
Radical initiator	$\geqslant C\cdot$	C_2H_4	$6.7 \times 10^6 \exp(-5000)/RT$	[115]

4 Chain Transfer Reactions

Using Ziegler-Natta catalysts, the termination of the polymer chain growth can proceed with the participation of a cocatalyst, a monomer, or hydrogen or via a β-hydride shift:

$$\text{⫴Mt—CH}_2\diagdown\diagup + \text{AlR}_3 \xrightarrow{k^{Al}_{tr}} \text{⫴Mt—R} + \text{R}_2\text{Al—CH}_2\diagdown\diagup \qquad (19)$$

$$\text{⫴Mt—CH}_2\text{—CHR}'\diagdown\diagup + \text{H}_2\text{C}=\text{CHR}'$$

$$\Big\downarrow k^M_{tr}$$

$$\text{⫴Mt—CH}_2\text{—CH}_2\text{R}' + \text{H}_2\text{C}=\text{CR}'\diagdown\diagup \qquad (20)$$

$$\text{⫴Mt—CH}_2\text{—CHR}'\diagdown\diagup \xrightarrow{k^s_{tr}} \text{⫴Mt—H} + \text{H}_2\text{C}=\text{CR}'\diagdown\diagup \qquad (21)$$

$$\text{⫴Mt—CH}_2\diagdown\diagup + \text{H}_2 \xrightarrow{k^H_{tr}} \text{⫴Mt—H} + \text{H}_3\text{C}\diagdown\diagup \qquad (22)$$

To analyze the transfer processes, in particular to estimate k^i_{tr} for the steady — state conditions when the rate of polymerization and the molecular mass of polymer do not vary with time, the following equation [2)] is used:

$$\frac{1}{\bar{P}_w} = \frac{k^M_{tr}}{k_p\omega} + \frac{1}{k_p\cdot\omega[M]}(k^s_{tr} + k^{Al}_{tr}[Al]^n + k^{H_2}_{tr}[H_2]^m) \qquad (VII)$$

For the nonsteady state polymerization the following equation [116)] holds:

$$\frac{1}{\bar{P}_w} = \frac{1}{\omega k_p[M]}\cdot\frac{V}{G} + \frac{k^M_{tr}}{\omega k_p} + \frac{1}{\omega[M]}\cdot\left(\frac{k^s_{tr}}{k_p} + \frac{k^{Al}_{tr}[Al]^n}{k_p} + \frac{k^H_{tr}[H_2]^m}{k_p}\right) \qquad (VIII)$$

In Eqs. (VII) and (VIII) $\omega = \bar{P}_w/\bar{P}_n$ is a coefficient characterizing the polymer polydispersity, $\bar{P}_w$ and $\bar{P}_n$ are the weight average and number average degree of polymerization, V and G the polymerization rate and polymer yield at a given moment of the reaction, [M], [Al], and [H₂] are the concentration of monomer, cocatalyst, and hydrogen, and n and m are the corresponding orders of transfer reaction with respect to cocatalyst and hydrogen concentrations.

4.1 Chain Transfer with Cocatalyst

The rate constant of this reaction (k^{Al}_{tr}) can be estimated according to Eq. (VII) from the data on the polymer molecular mass vs. cocatalyst concentration. However, k^{Al}_{tr} may also be determined directly by measuring the number of aluminium-polymer bonds (C_{Al}) formed in reaction (19). This determination may involve quenching of polymerization by alcohol with tritium-labelled hydroxyl [6,9,117)]. The parameter k_p, which is required for the calculation of k^{Al}_{tr} on the basis of C_{Al} data, should be determined independently using selective quenching agents. The k^{Al}_{tr} values for various catalytic systems are given in Tables 11 and 12. For TiCl₃ and TiCl₄/MgCl₂ these values are similar but strongly depend on the nature of monomer and AOC.

Table 11. Rate constants of chain transfer reactions in the olefin polymerization on catalysts containing titanium chlorides [29, 117–119]

Catalyst	Monomer	$k_{tr}^{AlEt_3}$	k_{tr}^M/ω	$k_{tr}^{H_2}/\omega$	$\tau^{a)} \times \omega$
		$1^{1/2} \times mol^{-1/2}$ $\times s^{-1}$	$1 \times mol^{-1} \times s^{-1}$	$1^{1/2} \times mol^{-1/2}$ $\times s^{-1}$	s
TiCl$_2$	C$_2$H$_4$	–	$0.8 \times 10^5\, e^{\frac{-9000}{RT}}$ 0.20 (80 °C)	–	12
TiCl$_3$ + AlEt$_3$	$\begin{cases} C_2H_4 \\ C_3H_6 \end{cases}$	2.3 (80 °C) 0.045 (70 °C)	0.084 (80 °C) 0.004 (70 °C)	2.3 (80 °C) 0.59 (70 °C)	25 160
TiCl$_4$/MgCl$_2$ + AlEt$_3$	C$_2$H$_4$	1.7 (80 °C)	0.22 (80 °C)	165$^{b)}$ (80 °C)	10

a) Average time of polymer chain growth calculated for a monomer concentration of 0.5 mol/l and the temperatures 80 °C (for ethylene) and 70 °C (for propylene);
b) Dimension of this value l/(mol × s)

Table 12. k_{tr}^{Al} of cocatalysts AlEt$_3$ and Al(i-Bu)$_3$ [14, 117]; catalyst: δ-TiCl$_3 \times 0.3$ AlCl$_3$

Cocatalyst	Monomer	$k_{tr}^{AlEt_3}$	$k_{tr}^{AlR_3}$
		$1^{1/2} \times mol^{-1/2} \times s^{-1}$	$1 \times mol^{-1} \times s^{-1}$
AlEt$_3$ Al(i-Bu)$_3$ $\}$	C$_2$H$_4$ (80 °C)	2.3 —	120$^{a)}$ 8.7
AlEt$_3$ Al(i-Bu)$_3$ $\}$	C$_3$H$_6$ (70 °C)	0.045 —	3.3$^{a)}$ 0.2

a) The data have been obtained from the expression $k_{tr}^{AlEt_3} \cdot K_D^{-1/2} \cdot K_D$: equilibrium constant of triethylaluminium dissociation

According to the adsorption mechanism of the reaction of chain transfer with OAC [117, 120)

$$\text{Ti}\overset{\square}{\underset{P}{\diagdown}} + AlR_3 \rightleftharpoons \text{Ti}\overset{R}{\underset{P}{\diagdown}}AlR_2 \longrightarrow \text{Ti}\overset{R}{\underset{\square}{\diagdown}} + PAlR_2 \qquad (23)$$

k_{tr}^{Al} depends on the nature of OAC and AC, in particular on the structure of the OAC alkyl group and the polymer chain (type of monomer). Due to steric reasons the change from linear to branched R and P, e.g. from Et to i-Bu and from C$_2$H$_4$ to C$_3$H$_6$, respectively, should prevent the adsorption of OAC on AC$_s$. The decrease of k_{tr}^{Al} is almost by two orders of magnitude lower for propylene than for ethylene polymerization [117, 121)] and by more than an order of magnitude lower for Al(i-Bu)$_3$ as compared with AlEt$_3$ (Table 12). Thus, the experimental order of k_{tr}^{Al} values calculated with the help of the AlEt$_3$ dissociation constant (Table 12)

$$k_{tr}^{AlEt_3}(C_2H_4) > k_{tr}^{Al(i\text{-}Bu)_3}(C_2H_4) \approx k_{tr}^{AlEt_3}(C_3H_6) > k_{tr}^{Al(i\text{-}Bu)_3}(C_3H_6) \qquad (VIII)$$

is in accord with the adsorption mechanism of chain transfer with a cocatalyst.

4.2 Chain Transfer with Monomer and Hydrogen

According to the data on the dependence of molecular mass vs. monomer and hydrogen concentration k_{tr}^M and $k_{tr}^{H_2}$ have been estimated by Eq. (VII) (Table 11). k_{tr}^M is slightly lower for titanium trichloride as compared with $TiCl_4/MgCl_2$ and $TiCl_2$, provided that the polydispersion coefficient ω is the same. However, in the presence of these catalysts polymers with different molecular mass distribution can be formed. This can be the reason for the observed difference in the average weight mass for a given catalyst.

For titanium trichloride k_{tr}^M, estimated in ref. [118] by taking into account k_P obtained by use of ^{14}CO, is practically independent of the cocatalyst nature. k_{tr}^M is strongly influenced by the monomer nature. For propylene polymerization k_{tr}^M is much lower (almost 20 times) than for ethylene polymerization (Table 11). k_P, however, differs more significantly (by two orders of magnitude). Thus, under similar reaction conditions, the polymer molecular mass is apparently lower in propylene than in ethylene polymerization. The rate constant of the chain transfer with hydrogen, $k_{tr}^{H_2}$, in the case of ethylene and propylene polymerization differs only by the factor four (Table 11); this is much lower than the differences in k_P. Hence, for a similar decrease of the molecular mass of polypropylene the hydrogen concentration should be much lower than in ethylene polymerization.

An essential difference is observed for the chain transfer with hydrogen in the polymerization on bulk $TiCl_3$ (the chain transfer is 0.5th order with respect to $[H_2]$) and on catalysts supported on $MgCl_2$ (first — order chain transfer with respect to $[H_2]$). This difference leads to higher values of the melt index of polyethylene prepared on the $TiCl_4/MgCl_2$ catalyst in the presence of H_2 in comparison with non-supported titanium chloride catalysts [119].

4.3 Contribution of Various Chain Transfer Processes

For all catalysts studied under the usual conditions of slurry polymerization of ethylene (75–80 °C, $[C_2H_4]$ = 0.1–0.6 mol/l) chain transfer with monomer predominates. This is directly illustrated by the independence of the polyethylene molecular mass of monomer concentration in the case of $TiCl_2$ [29] and $TiCl_4/MgCl_2$ + $AlEt_3$ [119]. The contribution of chain transfer with OAC in the case of $TiCl_4/MgCl_2$ is insignificant since the type and concentration of OAC do not influence the molecular mass of polyethylene [119]. Using titanium trichloride, the contribution of chain transfer with $AlEt_3$ increases with decreasing polymerization temperature, and at 50 °C chain transfer becomes the dominating process [117]. In propylene polymerization its contribution is several times lower [117].

Data on the significant role of the transfer reaction of the polymer chains with a monomer are in agreement with the number of vinyl and vinylidene groups in polyolefins [29, 118, 119].

In the above estimation of the contribution of various transfer processes to the polymer molecular mass, its steady-state value was taken into consideration. The molecular mass generally attains a steady-state value for 30–60 min (at 70–80 °C, $[M]$ = 0.1–1 mol/l) which at the first glance does not correspond to the short period of polymer chain growth $\bar{\tau}$ (Table 11). Apparently, this prolonged change of molecular mass with time can be attributed to additional transfer processes of the growing polymer chains with impurities present in the polymerization medium and with the products of the interaction of the components of the catalytic system. The concentration of these transfer agents seems to decrease with time due to their participation in various adsorption processes and reactions. All this leads to a slower increase of the average molecular mass of polymer at the beginning of polymerization. Nevertheless, the some polymer chains attain large mass close to the steady-state molecular mass of the polymer even for a short period of time (5 to 15 s) near $\bar{\tau}$ (Table 11) [34, 36, 38, 39].

5 Quantum-Chemical Interpretation of Olefin Interaction with the Active Center

In this section a brief review of quantum-chemical studies on the electron structure of the active center and the nature of the elementary steps of the chain propagation and transfer reactions for olefin polymerization is given.

5.1 Models of Active Center

The experimental data discussed in the previous sections correspond to the Cossee model of the monometallic active center [20, 21] (the octahedral complex Ti(III) is located on a lateral face of the titanium trichloride crystal)

The titanium ion contains an alkyl group and a chlorine vacancy. Other models of the AC have also been reported. Some of them do not include a metal — alkyl bond or vacant site [122–125]. Others are more complicated than Cossee's model, for example, Ti_2Cl_{10} clusters [126] or a bimetallic complex of Ti (IV),

(X=Cl, CH$_3$) [23, 26, 127], are considered as active centers.

    Armstrong was the first to note that the overall energy of the AC is diminished when the octahedral titanium environment is distorted to a trigonal bipyramid due to alkyl migration into a position which is intermediate between two octahedral vacancies [127]. This was confirmed by further calculations performed for monometallic [24, 128] or bimetallic [26] centers. Thus, the structure of active centers may formally be represented by a five-coordinated titanium ion without a vacant site:

(• : ion of Titanium)

According to Cossee [21] the back donation contributes a predominant part to the bonding between olefin and metal. However, this is not self evident for titanium ions in the oxidation states III and IV. Calculations of the π-complexes $Ti(III)X_5 \times C_2H_4$ (X=F, Cl) using the Slater-Johnson procedure [122, 123] show that the exchange of fluorine ions for chlorine ions results in a noticeable decrease of the role of dative bonds. However, even in the case of fluorine, the role of back donation in the formation to the metal-olefin bond is insignificant (less than 12%).

According to ref. [24] the Ti—C bond is mainly of covalent character with $Ti^{\delta+}$—$C^{\delta-}$ polarization. The length of the titanium-carbon bond, when passing from the methyl to the ethyl complex, changes insignificantly, and it can be supposed that the properties of this bond are independent of the chain length. In the formation of a π-complex an appreciable proportion of the electron density (≈ 0.25 e) is transferred from ethylene to the titanium ion.

5.2 Olefin Coordination Step

Olefin coordination with the active center, i.e. Ti(III) ion in the pentacoordinated environment with no formal vacant site, has to be accompanied by a reconstruction of its octahedral structure, due to the change of the alkyl position. This structural rearrangement requires energy which can be considered as the activation energy of the coordination step [127].

In refs. [24, 128], in terms of the CNDO method, the coordination of ethylene on a monometallic AC with various alkyl groups has been analyzed: CH_3 (complex A, reaction (24)), trans-C_3H_7 (complex C, reaction (25, 26)), and cis-C_3H_7 (complex F, reaction (27)–(29)). In all cases, the initial and final states of the AC in the process of ethylene coordination are optimized by the minimum of the overall energy. This process includes the construction of potential surfaces in two independent coordinates, according to the change of the position of alkyl and ethylene. Several possibilities of ethylene entering into the coordination sphere of Ti(III) ion have been calculated.

(A)   + C_2H_4 $\xrightarrow{E \approx 0}$ (B)   (24)

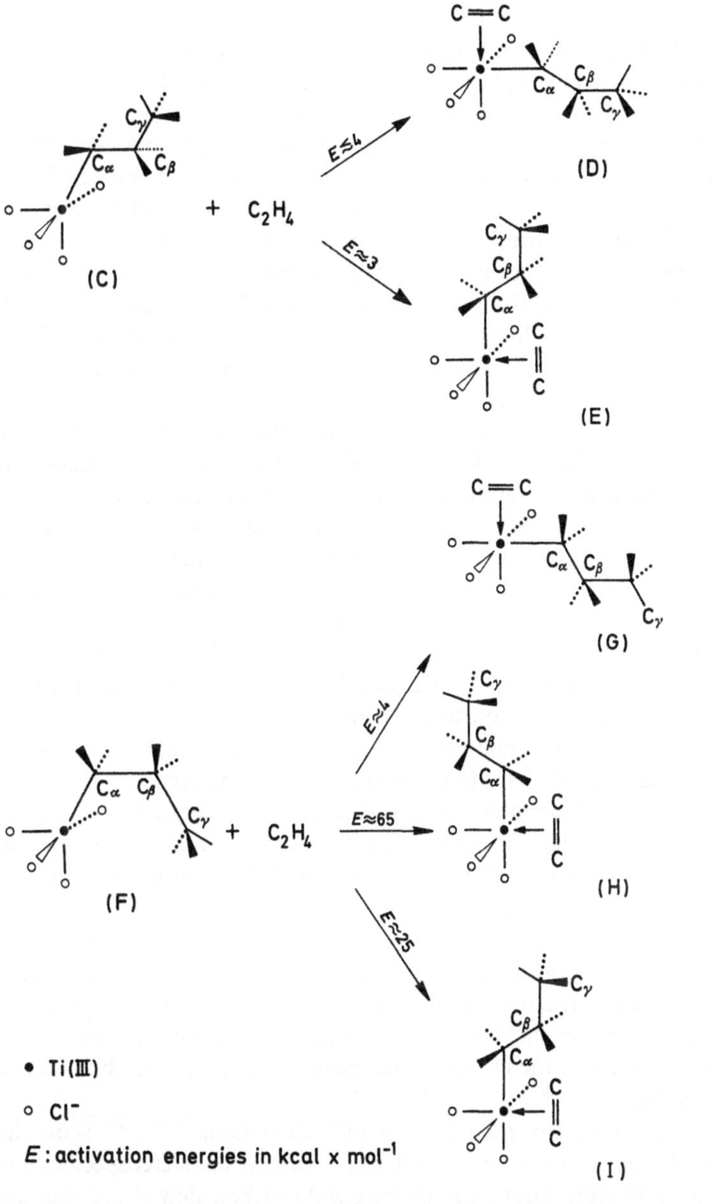

(C)   + C_2H_4

$E \leq 4$ (D)   (25)

$E \approx 3$ (E)   (26)

(G)   (27)

(F)   + C_2H_4

$E \approx 4$

$E \approx 65$ (H)   (28)

$E \approx 25$ (I)   (29)

• Ti(III)

○ Cl⁻

E: activation energies in kcal × mol⁻¹

When alkyl is CH_3 the energy of the system "active center + C_2H_4" during ethylene adsorption (reaction (24)) decreases without an appreciable energy barrier. A similar result is reported in ref. [26] in terms of an ab initio method. For C_3H_7 as the alkyl group the energy of coordination depends on the direction of the ethylene approach to the active center and on the conformation of the alkyl group.

In the case of a cis-alkyl group, olefin coordination from the side of the C_α carbon atom is energetically favored (reaction (27)). Ethylene adsorption from the side of the C_γ atom (reactions (28), (29)) requires a higher activation energy. The

high activation energies of routes (28) and (29) are due to the fact that because of the
more remote position of ethylene migration of the alkyl group and the change
of its structure from cis to trans should occur before energy compensation due to
ethylene coordination. Therefore, olefin adsorption occurs on the active centers
predominantly from the side of the C_α atom of the alkyl group with cis-conformation
(reaction (27)). After the cis-insertion of ethylene the complex G is transformed
into a complex with cis-conformation of the alkyl group which resembles the initial
complex F. Thus, adsorption of olefin on the same coordination site of titanium ions
before insertion becomes possible. This essential conclusion makes it unnecessary
to take into account the special step of the alkyl migration in the propagation
reaction [41] (reaction (30), step C) or to assume the existence on the $TiCl_3$ surface of
special disymmetric titanium ions as active centers [129] in order to explain the
isospecificity of the catalyst.

In the case of the active center, containing the transalkyl group (complex C, reaction (25), (26)) there is close probability of ethylene coordination to the both
coordination sites. Therefore, such center is non-stereospecific.

When the active center is located on the catalyst surface cis-trans conversion of
the alkyl may be hindered due to the steric effect of the surface. This results in the
stabilization of the active centers (complex F) which have the non-equivalence sites
for monomer coordination and can be stereospecific centers. If the ligand environment of the active center permits rapid conversion of the cis-alkyl group (complex F)
to the trans-alkyl (complex C) the coordination sites become equivalent; in this case,
the active center will be nonstereospecific.

In the active center containing $Ti-CH_3$ bond the coordination sites are equivalent,
and the addition of olefin is nonstereospecific. However, the insertion of the first
monomer molecule creates non-equivalence of the coordination sites at the active
center, thus, the addition of the second' molecule becomes stereospecific. This may
explain the experimental results of Zambelli et al. [66] on the influence of the size of
an alkyl ligand on the stereospecificity of monomer addition (Table 5). The conclusion
on the influence of the alkyl group on the stereospecificity was drawn also by
Corradini et al. [69] on the basis of calculations on the non-bonding interactions for
the active center located on the surface of $TiCl_3$, propylene being considered as
monomer. However, in this case, the propylene π-complexes were studied without
taking into account the change of the position of the $Ti-C_\alpha$ bond during the coordination step.

Thus, recent experimental data [66] and calculations [24, 69, 128] show the important
role of the end of the growing polymer chain on the stereospecificity of the olefin
addition. This role is due to the influence of the carbon atom (C_β, C_γ) of the main chain
on the isospecificity of the olefin addition and not to the effect of the substituent of the
last monomeric unit as in the case of the syndiospecific addition. The growing polymer
chain is one of ,the ligand which, together with other ligands of titanium ions,
determines the chirality of the active center.

5.3 Olefin Insertion Step

Cossee [20, 21] was the first to analyze theoretically olefin insertion into metal-alkyl
bonds (reaction (30), step B) assuming that it was the rate-determining step of the

propagation reaction:

$$(30)$$

After cis-addition of a monomer the geometry of an active center is reconstructed with a change of the positions of the vacancy and the alkyl. From the estimation of the contribution of the atomic orbitals to the molecular orbital localized at the Ti—C bond, it was concluded that the metal-alkyl bond is preserved during the reaction route. Cossee [21] regarded the similar reactions to the concerted-type reactions which are characterized by synchronous redistribution of the electron density between dissociated and newly formed bonds which accounts for the low activation energies of these reactions.

On the basis of their calculations Cossee [21] and then Begley and Penella [130] concluded that the activation energy of the insertion step depends on the energy ΔE of the transfer from the highest occupied orbital σ_R, attributed to the metal-alkyl bond to the partly occupied or vacant d_π-orbital of the metal ion. However, this conclusion was not confirmed by CNDO [23, 24, 127] or ab initio [26] calculations. The value of ΔE strongly depends on the lengths of metal-alkyl and metal-ethylene bonds [24]. It was therefore important to determine the equilibrium lengths of these bonds from the minimum of the total energy. In refs. [21, 130] this factor has not been taken into account.

A CNDO analysis of the insertion step of coordinated ethylene for the bimetallic active center [23, 127] showed that this step proceeds with an activation energy which is lower than the heat of formation of the π-complex. A low activation energy ≈ 1 kcal/mol) for the insertion step was calculated also for the monometallic active center [24], a mutual influence of Ti-ethylene and Ti-alkyl being taken into account.

The possibility of ethylene insertion into the Ti—C bond without the formation of an intermediate π-complex (one-step mechanism) has also been calculated [24]. The activation energy of the one-step process is much higher than that of the two = step process. CNDO calculations have revealed also that the insertion of ethylene into the Ti—C bond of the tetrahedral complex Cl_3TiCH_3 requires a very high activation energy (76 kcal/mol) [131]. This high energy barrier may be due to the structure of the $Cl_3TiCH_3 \cdot (C_2H_4)$ complex for which an optimum distance of 3.5 Å compared with 2.3 Å for the corresponding π-complexes [22, 24, 26] between the titanium ion and ethylene was calculated. Thus, the ethylene insertion studied in ref. [131] may be considered as an intermediate between a one-step and a two-step mechanism.

The activation energy of the insertion of coordinated ethylene estimated by the ab initio method was found to be ≈ 15 kcal/mol [26]. Despite the application of a more advanced calculation technique these results are less compatible with the experimental data on solid titanium chloride-based catalysts, when the activation energy of the propagation step is 3–6 kcal/mol (Table 10). Probably, this incompatability is due to the model used in ref. [26] which describes the AC as a bimetallic complex Cl_3TiCH_3 with $Al(CH_3)_3$. However, it is important to note that the calculations performed by means of the nonempirical method confirm the concept implying that in the active center the alkyl group occupies an intermediate position between the octahedral sites and that in olefin coordination the AC structure is reconstructed.

5.4 Chain Transfer with Monomer

This reaction involves transfer of a hydrogen atom from the β-carbon of the alkyl chain to the coordinated olefin:

(31)

CNDO calculations of this transfer reaction have revealed [23] that it does not require a significant activation energy and that it competes with the insertion of coordinated ethylene into the Ti—C bond. However, the rates of the propagation and the transfer reaction with a monomer differ greatly, leading to the formation of high molecular weight polyolefins (the degree of polymerization is 10^4–10^6). Apparently, this difference is explained by the fact that the formation of π-complexes with different relative arrangement of alkyl and olefin is necessary for these competetive reactions. Propagation proceeds via a π-complex formed in the ethylene coordination on the site of C_α atom (reactions (25), (27)). This is unfavourable for a β-hydride transfer due to steric reasons. The transfer of the polymer chain with a monomer proceeds via other π-complexes (E), (I) forming at the ethylene adsorption on the side of the C_β and C_γ atoms (reactions (26), (29)). The high degree of polymerization implies that the activation energy of ethylene adsorption from the C_α site should be lower than from the other coordination site. This corresponds to the calculated activation energies of ethylene coordination, particularly in the case of an AC with a cis-structure of the alkyl group (reactions (27) and (29)). Thus, the formation of high molecular weight and isotactic polymers is best described by a mechanism which assumes an adsorption of the olefin mainly on the same site of the coordination sphere of the titanium ion.

5.5 Olefin Interaction with Organoaluminium Compounds

The catalytic oligomerization of olefins in the presence of OAC and the olefin polymerization in the presence of transition metals are based on similar olefin insertions into the metal-carbon and metal-hydrogen bonds (see Section 3.2). However, in organoaluminium compounds, the structure of the active center is defined more simply and more reliably. Data on its coordination state, thermodynamic and kinetic parameters have been reported (e.g. Table 13).

Table 13. Bond energies (D) of metal-carbon or metal-hydrogen bonds and activation energies (E) of the insertion of ethylene into these bonds

Catalyst	Active bond	D	E
		$kcal \times mol^{-1}$	$kcal \times mol^{-1}$
$Al(CH_3)_3$	Al—C	64.5	22.5 [110]
$(i-C_4H_9)_2AlH$	Al—H	68	4.9 [112]
$TiCl_3$-AlR_3	Ti—C	44–66[a]	3.0[55]

[a] The range given corresponds to the bond energy of the Ti-C bond in different organotitanium compounds [132,133]

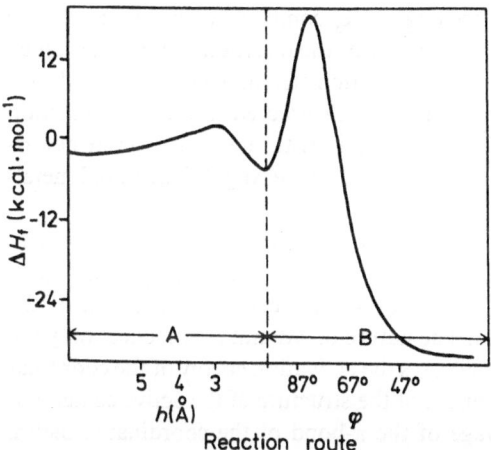

Fig. 3. Data profile for the interaction of ethylene with $Al(CH_3)_3$ [25]. Part A: coordination step. As reaction coordinate the distance h between aluminium and ethylene is considered. Part B: insertion step, the angle φ between the Al—CH_3 bond and the Al—$\overset{C}{\underset{C}{\|}}$ bond representing the reaction coordinate.

The addition of ethylene to Al—C and Al—H bonds in $Al(CH_3)_3$ and AlH_3 molecules was studied by means of the MINDO-3 method [25]. The obtained data indicate that the insertion of ethylene into Al—C and Al—H bonds proceeds via an intermediate π-complex (Fig. 3).

For both types of bonds the activation energies of the coordination steps are fairly close (Table 14) whereas those of the insertion steps differ greatly since the

Table 14. Calculated activation energies (E) of the addition reaction of olefins into Al—C and Al—H bonds [25]

Active center	Active bond	Olefin	E (kcal/mol)		
			coordination	insertion	overall[a]
AlH_3	Al—H	Ethylene	2.8	6.4	2.8
$(CH_3)_2AlH$	Al—H	Propylene	5.7	4.9	5.7
$(CH_3)_2AlCH_3$	Al—CH$_3$	Ethylene	5.1	24.7	21.4

[a] Equal to the activation energy of the coordination step in the case of AlH_3 and $(CH_3)_2AlH$ and to the difference between the activation energy of the insertion step and heat of formation of the π-complex in the case of $Al(CH_3)_3$

insertion of ethylene into the Al—C bond requires a higher activation energy. Thus, for the olefin addition to the Al—H bond the coordination step may be the rate-determining one, whereas for the Al—C bond the insertion step is the rate-determining step.

The calculated activation energies of the ehylene addition to Al—C and Al—H (Table 14) bonds agree well with the experimental data (Table 13). According to the calculated data, the rearrangement of the structure of the active center occurs during the coordination of ethylene. The flat trigonal AlR_3 complex is converted into the tetrahedron of $(C_2H_4)AlR_3$. The activation energy of the insertion step is determined by the rupture of the olefin double bond. The insertion step may be represented by a sequence of rearrangements of chemical bonds which proceed in the same reaction act. First, cleavage of the π-bond of the olefin occurs with the formation of new σ-bonds, the evolved energy compensating the subsequent cleavage of the initial metal-ligand σ-bond.

Thus, the addition of olefin to the Mt—L bond (where L is alkyl or hydrogen), both for transition and non-transition metals (Mt), proceeds according to the same two-step mechanism via intermediate π-complexes with further cis-insertion of the coordinated olefin into the Mt—L bond. The nature of Mt and L influences only the activation energies of separate steps. In all cases the activation energy of the coordination step is determined by the rearrangement of the structure of the active center and that of the insertion step by the cleavage of the π-bond of the coordinated olefin. Hence, it is clear that the activation energies of the olefin insertion into the Mt—C and Mt—H bonds are not compatible with the energies of these bonds. This is illustrated by the experimental data listed in Table 13. Thus, attempts [125] to correlate the reactivity of metal-ligand bonds with their thermodynamic parameters have no correct basis.

6 Conclusion

Using well developed methods for the determination of the number of active centers and propagation rate constants, it became possible to trace the effect of catalyst composition and polymerization conditions on these parameters. The data obtained generally agree with the concept of the monometallic composition of the AC the reactivity of which does not depend on the nature of the organometallic cocatalyst. This conclusion is also confirmed by the results of an NMR study of the polypropylene structure.

Isospecific polymerization is characterized by a high regioselectivity. Stereocontrol of olefin addition is due to the ligand environment of the transition metal of the AC, including not only inorganic ligands but also β- and γ-carbon atoms of the growing polymer chain. In syndiospecific polymerization stereocontrol of olefin addition is determined by the influence of the substituent of the last monomeric unit of the growing end of the polymer chain.

The activity of the catalyst in the olefin polymerization changes a wide range mainly with the variation of the number of ACs. This number depends on the nature and crystalline structure of the transition metal compound, the presence of a support and catalyst modifier, the nature of a cocatalyst, and the polymerization conditions. The most pronounced increase of the number of active centers (up to nearly half of the total content of titanium in the catalyst) is achieved by supporting titanium chlorides on anhydrous highly dispersed magnesium chloride. Catalysts of this type show the highest activity amongst all known catalytic systems used for olefin polymerization.

Kinetic data and quantum-mechanical calculations confirm the two-step mechanism of propagation-coordination of the olefin on the metal ion with subsequent cis-insertion into the metal-carbon σ-bond. Coordination of the olefin is the rate-determining step for most of the catalytic systems. This mechanism is valid for polymerization catalyzed by transition metal complexes as well as for the olefin oligomerization in the presence of individual organoaluminium compounds. The difference in these two processes is due to the different rate constants of the various stages of these processes. The addition of the olefin to the transition metal-carbon bond proceeds with an activation energy which is considerably lower than that required for the addition to the aluminium-carbon bond. Based on quantum-chemical calculations the conclusion was drawn that the activation energy of the coordination step is related to the energy necessary for the structural rearrangement of the active center, and the activation energy of the insertion step to the dissociation of the π-bond of the olefin.

Further progress in the elucidation of the mechanism of catalytic polymerization of olefins may be expected from a wider application of physical methods to the identification of active centers in the polymerization and to the study of their structure and reactivity.

7 References

1. Ziegler, K. et al.: Angew. Chem. 67, 541 (1955)
2. Natta, G., Pasquon, I.: Adv. Catal. 11, 1 (1959)
3. Boor, J.: Macromol. Rev. 2, 115 (1967)
4. Boor, J.: Ziegler-Natta Catalysts and Polymerizations, Academic Press, New York 1979
5. Yermakov, Yu. I., Zakharov, V. A.: Adv. Catal. 24, 173 (1975)
6. Schnecko, H., Kern, W.: Chem.-Ztg., Chem. Appar. 94, 229 (1970)
7. Yermakov, Yu. I., Zakharov, V. A.: Usp. Khim. 41, 377 (1972)
8. Feldman, G. F., Perry, E.: J. Polym. Sci. 46, 217 (1960)

9. Tait, P. J. T., in: Coordination Polymerization (ed.) Chien, J. C. W., p. 155, Academic Press, New York-San Francisco-London 1975
10. Yermakov, Yu. I., Zakharov, V. A., Bukatov, G. D.: Proc. V Int. Congr. Catal., vol. 1, p. 399, Amsterdam-New York 1973
11. Zakharov, V. A. et al.: React. Kinet. Catal. Lett. *1*, 247 (1974)
12. Yermakov, Yu. I., Zakharov, V. A., in: Coordination Polymerization (ed.) Chien, J. C. W., p. 91, Academic Press, New York-San Francisco-London 1975
13. Bukatov, G. D. et al.: Makromol. Chem. *179*, 2097 (1978)
14. Zakharov, V. A. et al.: Kinet. i Katal. *18*, 848 (1977)
15. Bukatov, G. D., Zakharov, V. A., Yermakov, Yu. I.: Kinet. i Katal. *16*, 645 (1975)
16. Ziegler, K., in: Organometallic Chemistry (ed.) Zeiss, H., Reinhold Publ. Corp., N.Y., Chapman & Hall, Ltd., London 1960
17. Bovey, F. A.: High Resolution of Macromolecules, Academic Press, New York 1972
18. Sivaram, S.: Ind. Eng. Chem., Prod. Res. Dev. *18*, 121 (1977)
19. Zakharov, V. A., Yermakov, Yu. I., in: Catalysts and Catalytical Processes, p. 135, Institute of Catalysis, Novosibirsk 1977
20. Cossee, P.: J. Catal. *3*, 80 (1964)
21. Cossee, P., Ros, P., Schachtschneider, J. H.: Proc. IVth Intern Congr. Catal., p. 237 Moscow, 1968, Academiai Kiado, Budapest 1971
22. Armstrong, D. R., Fortune, F., Perkins, P. G.: J. Catal. *42*, 435 (1976)
23. Novaro, O., Chow, S., Magnonat, P.: J. Catal. *41*, 91 (1976); *42*, 131 (1976)
24. Avdeev, V. I. et al.: Zh. Strukt. Khim. *18*, 525 (1977)
25. Zakharov, I. I., Zakharov, V. A.: J. Molec. Catal. *14*, 171 (1982)
26. Novaro, O. et al.: J. Chem. Phys. *68*, 2337 (1978)
27. Kendlin, J., Taylor, K., Tompson, D.: Reactions of Coordination Compounds of Transition Metal, Elsevier Publ. Co., Amsterdam-London-New York 1968
28. Warzelhan, V., Burger, T. F.: IUPAC, Intern. Symp. Macromol., Florence, 1980, Preprints, V. 2, p. 13
29. Zakharov, V. A. et al.: Kinet. i Katal. *16*, 417 (1975)
30. Mejzlik, J., Lesna, M.: Makromol. Chem. *178*, 261 (1977)
31. Chumaevskii, N. B. et al.: Makromol. Chem. *177*, 747 (1976)
32. Zakharov, V. A. et al.: Dokl. AN SSSR *207*, 857 (1972)
33. Zakharov, V. A. et al.: React. Kinet. Catal. Lett. *2*, 329 (1975)
34. Shepelev, S. N. et al.: Kinet. i Katal. *22*, 258 (1981)
35. Giannini, U.: Makromol. Chem., Suppl. *5*, 216 (1981)
36. Bukatov, G. D. et al.: Makromol. Chem., *183*, 2657 (1982)
37. Coover, H. W. et al.: J. Polym. Sci. *58*, 2583 (1966)
38. Böhm, L. L.: Polymer *19*, 562 (1978)
39. Suzuki, E. et al.: Makromol. Chem. *180*, 2235 (1979)
40. Natta, G. et al.: Chim. Ind. (Milano) *40*, 717, 896 (1958)
41. Arlman, E. J., Cossee, P.: J. Catal. *3*, 89, 99 (1964)
42. Arlman, E. J.: J. Catal. *5*, 178 (1966)
43. Rodriguez, L. A. M., van Looy, H. M.: J. Polym. Sci., Part A-1, *4*, 1971 (1966)
44. Buls, V. M., Higgins, T. L.: J. Polym. Sci., Part A-1, *8*, 1025 (1970)
45. Carradine, W. R., Rase, H. F.: J. Appl. Polym. Sci. *15*, 889 (1971)
46. Munoz-Escalona, A., Parada, A.: Polymer *20*, 474 (1979)
47. Wilshinsky, Z. Wm., Looney, R. W., Tornqvist, G. M.: J. Catal. *28*, 351 (1973)
48. Zakharov, V. A. et al.: Kinet. i Katal. *16*, 1184 (1975)
49. Vermel, E. E. et al.: Kinet. i Katal. *22*, 480 (1981)
50. Zakharov, V. A. et al.: Kinet. i Katal. *15*, 446 (1974)
51. Maksimov, M. G. et al.: Kinet. i Katal. *15*, 738 (1974)
52. Zakharov, V. A.: Kinet. i Katal. *21*, 892 (1980)
53. Makhtarulin, S. I. et al.: React, Kinet. Catal. Lett. *9*, 269 (1978)
54. Vermel, E. E. et al.: Vysokomol. Soed., Ser. A, *22*, 22 (1980)
55. Zakharov, V. A. et al.: Makromol. Chem. *177*, 763 (1976)
56. Zakharov, V. A. et al.: Makromol. Chem. *178*, 967 (1977)
57. Natta, G. et al.: Chim. Ind. (Milano) *48*, 1298 (1966)
58. Zambelli, A., Tosi, C.: Adv. Polym. Sci. *15*, 31 (1974)

59. Asakura, T. et al.: Makromol. Chem. *178*, 791 (1977)
60. Zambelli, A., Bajo, G., Rigamonti, E.: Makromol. Chem. *179*, 1249 (1978)
61. Zambelli, A., Locatelli, P., Rigamonti, E.: Macromolecules *12*, 156 (1979)
62. Zambelli, A. et al.: Macromolecules *13*, 798 (1980)
63. Doi, Y.: Macromolecules *12*, 248 (1979)
64. Zambelli, A. et al.: Macromolecules *4*, 475 (1971)
65. Zambelli, A., Allegra, G.: Macromolecules *13*, 42 (1980)
66. Zambelli, A. et al.: Macromolecules, *15*, 211 (1982)
67. Doi, Y. et al.: Makromol. Chem. *176*, 2159 (1975)
68. Zambelli, A., Sacchi, M. C., Locatelli, P.: Internat. Symposium on "Transition metal catalysed polymerizations: unsolved problems", Midland, August 1981
69. Corradini, P. et al.: Eur. Polym. J. *15*, 1133 (1979); *16*, 835 (1980)
70. Wolfsgruber, C. et al.: Makromol. Chem. *176*, 2765 (1975)
71. Bukatov, G. D. et al.: Makromol. Chem. *179*, 2093 (1978)
72. Doi, Y., Suzuki, E., Keii, T. K.: Makromol. Chem., Rapid Commun. *2*, 293 (1981)
73. Jordan, D. O., Hoeg, D. F., in: Stereochemistry of Macromolecules (ed.) Ketley, A. D., Vol. 1, M. Dekker, New York 1967
74. Keii, T.: Kinetics of Ziegler-Natta Polymerization, Kodansha, Tokyo 1972
75. Chirkov, N. M., Matkovskii, P. E., Dyachkovskii, F. S.: Polymerization on Complex Organometallic Catalysts, Khim., Moscow 1976
76. Keii, T., in: Coordination Polymerization (ed.) Chien, J. C. W., p. 263, Acad. Press, N.Y. 1975
77. Kohara, T. et al.: Makromol. Chem. *180*, 2139 (1979)
78. Zakharov, V. A., Bukatov, G. D., Yermakov, Yu. I.: Makromol. Chem. *176*, 1959 (1975)
79. Natta, G., Danusso, F., Sianesi, D.: Makromol. Chem. *30*, 238 (1959)
80. Zambelli, A. et al.: Makromol. Chem. *112*, 160 (1968)
81. Zavorokhin, N. D. et al.: Dokl. AN SSSR *215*, 593 (1974)
82. Bokareva, N. V., Tsvetkova, V. I., Dyachkovskii, F. S.: Vysokomol. Soed., Ser. A, *20*, 2707 (1978)
83. Maksimov, N. G. et al.: Teor. i Exper. Khim. *14*, 53 (1978)
84. Maksimov, N. G. et al.: J. Mol. Catal. *4*, 167 (1978)
85. Ivin, K. J. et al.: J. Chem. Soc., Chem. Commun. *1978*, 604
86. Dolgoplosk, V. A. et al.: Vysokomol. Soed., Ser. A, *19*, 2464 (1977) (see refs. therein)
87. Evitt, E. R., Bergman, R. G.: J. Am. Chem. Soc. *101*, 3973 (1979)
88. Locatelli, P. et al.: IUPAC, Internat. Symp. Macromol., Florence, 1980, Preprints, vol. 2, p. 28
89. Casey, C. P.: Macromolecules *14*, 464 (1981)
90. McKinney, R. J.: J. Chem. Soc., Chem. Commun. *1980*, 490
91. Zakharov, V. A., Yermakov, Yu. I.: Catal. Rev.-Sci. Eng. *19*, 67 (1979)
92. Ziegler, K. et al.: Liebigs Ann. Chem. *629*, 53 (1960)
93. Ziegler, K., Hoberg, H.: Chem. Ber. *93*, 2944 (1960)
94. Hay, J. N., Jones, G. R., Robb, J. C.: J. Organomet. Chem. *15*, 295 (1968)
95. Hay, J. N., Hooper, P. G., Robb, J. C.: Trans. Farad. Soc. *65*, 1365 (1969)
96. Hay, J. N., Hooper, P. G., Robb, J. C.: Trans. Farad. Soc. *66*, 2800 (1970)
97. Allen, P. E. M., Byers, A. E.: Trans. Farad. Soc. *67*, 1718 (1971)
98. Allen, P. E. M. et al.: Trans. Farad. Soc. *63*, 1636 (1967)
99. Allen, P. E. M., Byers, A. E., Lough, R. M.: J. Chem. Soc., Dalton Trans. *4*, 479 (1972)
100. Allen, P. E. M., Lough, R. M.: J. Chem. Soc., Farad. Trans. 1, *69*, 849 (1973)
101. Smith, M. B.: J. Phys. Chem. *71*, 364 (1967)
102. Hata, G.: Chem. Commun. *1968*, 7
103. Dawies, J. St., Oliver, J. P., Smart, J. B.: J. Organomet. Chem. *44*, C32 (1972)
104. Dobzine, T. W., Oliver, J. R.: J. Am. Chem. Soc. *96*, 1737 (1974)
105. Hay, J. N., Hooper, P. G., Robb, J. C.: Trans. Faraday Soc. *66*, 2045 (1970)
106. Hay, J. N., Hooper, P. G., Robb, J. C.: J. Organomet. Chem. *28*, 193 (1971)
107. Smith, M. B.: J. Organomet. Chem. *46*, 211 (1972)
108. Henrikson, C. H., Eyman, D. P.: Inorg. Chem. *6*, 1461 (1967)
109. Egger, K. W.: Trans. Faraday Soc. *67*, 2638 (1971)
110. Egger, K. W., Cocks, A. T.: J. Am. Chem. Soc. *94*, 1810 (1972)
111. Egger, K. W.: J. Am. Chem. Soc. *91*, 2867 (1969)

112. Egger, K. W.: J. Chem. Soc., Faraday Trans. 1, *68*, 423 (1972)
113. Egger, K. W.: J. Chem. Soc. A *1973*, 3603
114. Cocks, A. T., Egger, K. W.: J. Chem. Soc. A *1971*, 3606
115. Bagdasaryan, Kh. S.: Theory of Radical Polymerization, Nauka, Moscow 1966
116. Ivanov, L. P., Yermakov, Yu. I., Gelbshtein, A. I.: Vysokomol. Soed., Ser. A, *9*, 2422 (1967)
117. Bukatov, G. D. et al.: Makromol. Chem. *178*, 953 (1977)
118. Zakharov, V. A. et al.: React. Kinet. Catal. Lett. *5*, 429 (1976)
119. Zakharov, V. A. et al.: Vysokomol. Soed., Ser. A, *21*, 496 (1979)
120. Cossee, P.: Trans. Farad. Soc. *58*, 1926 (1962)
121. Jung, K. A., Schnecko, H.: Makromol. Chem. *154*, 227 (1972)
122. Ros, P.: Chem. Biochem. Reactiv. Proc. Internat. Symp., p. 207, Jerusalem 1974
123. Rösch, N., Jonson, K. H.: J. Mol. Catal. *1*, 395 (1976)
124. Armstrong, D. R., Perkins. P. G., Stewart, J. J. P.: Rev. Roumaine Chim. *20*, 177 (1975)
125. Lvovskii, V. E., Erusalimskii, G. B.: Coord. Khim. *2*, 1221 (1976)
126. Armstrong, D. R., Perkins, P. G., Stewart, J. J. P.: Rev. Roumaine Chim. *19*, 1695 (1974)
127. Armstrong, D. R., Perkins, P. G., Stewart, J. J. P.: J. Chem. Soc., Dalton Trans. *1972*, 1972
128. Zakharov, I. I. et al., in: Complex Organometallic Catalysts of Olefin Polymerization, p. 19, Institute Chem. Phys. AN SSSR, Chernogolovka 1980
129. Allegra, G.: Makromol. Chem. *145*, 235 (1971)
130. Begley, J. W., Ponella, F.: J. Catal. *8*, 203 (1967)
131. Cassoux, P., Crasnier, F., Laberre, J. F.: J. Organomet. Chem. *165*, 303 (1979)
132. Telnyi, V. I. et al.: Dokl. AN SSSR *174*, 1374 (1967)
133. Lappert, M. F., Patil, D. S., Pedley, J. B.: J. Chem. Soc., Chem. Commun. *1975*, 830
134. Pino, P., Mühlhaupt, R.: Angew. Chem., Int. Ed. Engl. *92*, 857 (1980)
135. Galli, P., Luchiani, L., Cecchin, G.: Angew. Makromol. Chem. *94*, 63 (1981)
136. Bier, G.: Polymer Bulletin, *7*, 177 (1982)
137. Bukatov, G. D., Zakharov, V. A., Yermakov, Yu. I.: Polymer Bulletin, (1982), in press.

W. Kern (editor)
Received October 1st, 1982

Control of Molecular-Weight Distribution in Polyolefins Synthesized with Ziegler-Natta Catalytic Systems

Umberto Zucchini and Giuliano Cecchin
Montepolimeri S.p.A., Centro Ricerche G. Natta, Piazzale G. Donegani,
44100 Ferrara, Italia

In spite of the enormous number of papers published and patents issued on Ziegler-Natta catalysis, it does not exist a review on the methods of polyolefins molecular weight distribution (MWD) control. In the present article we shall review scientific and patent literature on this argument.

After a short comment on the speculative and industrial importance of polyolefins MWD, an outlook is given on the theories of the origins of the wide MWD usually shown by polyolefins. Subsequently, a comprehensive critical survey of the possibilities of MWD control, based mainly on the type of catalytic system and on polymerization parameters, is discussed. Finally, some considerations on MWD control in industrial processes are given and a rationalized collection of the most significant patents for polyethylene MWD control since 1968 is presented.

In conclusion, the theories based on the plurality of the catalytic active species appear more convincing than those based only on physical phenomena in explaining MWD. Together with some general principles, only a better knowlegde of number and types of polymerization centres and of the relevant kinetic constants could lead to a more effective MWD control. This should represent one of the future trends of research and development in Ziegler-Natta catalysis.

1 Introduction

The revolutionary discoveries by Ziegler and Natta, relating to the low pressure polymerization, respectively, of ethylene and of propylene and other α-olefins onto the previously unknown crystalline polymers, opened a new era in science and technology. Since then, remarkable progress has been made in the fields of coordination catalysis, macromolecular science and stereochemistry. With the discovery and development of the "new generation" catalytic systems[1] for polyethylene in the late 1960's, and more recently for polypropylene, enormous progress was made in terms of polymerization process as to economics and product quality [1]. Further process simplification and, above all, ever more accurate product quality control by "taylor made" catalytic systems is the aim of the 1980's.

To this end, knowledge of the principles which govern polymer molecular weight distribution (MWD) and the methods by which it can be controlled, is of extreme importance both from a theoretical and industrial point of view.

Although in recent years knowledge regarding Ziegler-Natta catalysis has been reported in various publications and reviews [2-5, 250], MWD regulation has been dealt with [6-13, 240], but not as thoroughly as it should be for its practical and theoretical relevance.

A great quantity of information about polyolefins MWD and the possibilities for its regulation is available from literature, patents and from practical experience, both in the laboratory and in industrial production. However, even when the cases are limited to the most simple and common methods, the theoretical interpretation of the results is still uncertain and inconclusive. Indeed one often finds in the literature that the principles for obtaining broad MWDs are not yet exactly known [8] or, as also Boor states [6]: "Beyond recognizing that soluble catalysts lead to narrower MWD than obtained with heterogeneous catalysts, not much else of significant importance has been reported on the control of MWD".

In the present review we have critically collected the available information without, however, pretending either to be exhaustive or to provide definitive explanations for all the problem involved with such a wide variety of polymerization processes [14, 15] and catalytic systems [6, 9, 16, 17].

Thus, the literature from the beginning until the first half of 1982 was examined, paying also attention, expecially in the case of polyethylene, even to the patent literature available from 1968, that is, since the discovery of the so-called "second generation" catalytic systems. As the most important production processes involve the polymerization of ethylene and propylene, the discussion will be limited to these homopolymers. With regard to polymer MWD regulation, the most significant methods based on the choice of the catalyst system and the polymerization conditions will be examined.

Other methods for polyolefins MWD regulation, which will not be dealt with here, are *post-treatments* on the polymer such as blending of products with different molecular weights [18], and thermal, catalytic or mechanical visbreaking of the polymer itself [19-22].

---

[1] In this review the catalytic system is the combination between a transition metal compound (catalyst) and a metal alkyl (cocatalyst)

2 Significance and Importance of MWD

In Ziegler-Natta polymerization, at any given moment during the growth of a macromolecule, many possibilities for its evolution are encountered. The chain end can add a new monomeric unit, or may terminate spontaneously or by a transfer reaction, or because the active center, upon which it grows, has been deactivated. Therefore, as well known, every polymer is generally composed of macromolecules with different degree of polymerization (length of the chain).

These lengths are distributed according to a function which depends on polymerization mechanism, and the resulting polydispersity state of the polymer, with regard to its macromolecules (i.e. MWD), can be described by a curve plotting number fraction, or weight fraction versus a parameter representing molecular size (chain length, molecular weight, etc.).

The calculation of polymer MWD essentially involves the evaluation of N_r, number of polymeric chains containing r-monomeric units, as a function of r.

The breadth of MWD can be evaluated by a ratio between different molecular weight averages. The most frequently quoted is the polydispersity index Q, defined as $\bar{M}_w/\bar{M}_n$, where $\bar{M}_w$ and $\bar{M}_n$ are the weight and the number average molecular weight respectively. Other less used functions are the so-called "inhomogeneity" proposed by Schulz [255], expressed as $(\bar{M}_w/\bar{M}_n) - 1$, or those functions which utilize ratios between higher averages.

Knowledge about average molecular weights only gives partial indications as to the nature of the polymer, while analytic methods which give their distribution function, provide a more complete description.

2.1 Principal Methods of MWD Determination

In this section we only intend to outline the main problems which are ecountered in polyolefins MWD determination, without a critical discussion on this complex matter. There are several techniques for determining the main MWD averages [34, 36, 256]. They all utilize the properties of polymer solutions, some giving direct results, others requiring a calibration [257].

a) *Fractional precipitation*

Polymer fractionation is carried out on the basis of the decreasing solubility of polymer with increasing molecular weight. In the fractional precipitation method, the polymer is dissolved in a solvent and a non-solvent is added gradually at constant temperature.

An inherent problem with the fractional precipitation method is that each fraction may contain appreciable concentrations of lower molecular species. Furthermore fractionation of higher molecular weight species is carried out at concentrations higher than the those for lower molecular weight, with consequent increasing of separation efficiency with decreasing molecular weight. However, this method is very time consuming.

b) *Column chromatography*

The most important techniques of the chromatography principle are column elution based on a solvent — non solvent gradient, and column elution with a superimposed temperature gradient.

The *solvent gradient* usually consists of a change in composition of a binary solvent with time at constant temperature. The *temperature gradient* usually consists of a change of temperature to increase the solving power of the solvent. Efficiency of fractionation decreases with increasing sample size, and is inversely proportional, also from a theoretical point of view, to molecular weight. Actually for crystallin polymers, as in polypropylene, the relationship solubility — molecular weight is affected also by the stereoregularity of the macromolecule.

c) *Gel permeation chromatography (G.P.C.)*

The separation of polymer macromolecules by G.P.C. is based upon differences in hydrodynamic volume, that is their actual size in solution, closely related to molecular weight. Separation is accomplished by injecting polymer solution into a continuously flowing stream of solvent which passes through porous, tiny, rigid gel particles. Low molecular weight macromolecules take longer time to emerge than the larger ones, due to their greater penetration into the particle pores. In order to obtain absolute data of molecular weight and MWD it is necessary to calibrate G.P.C. outputs; the best way is the use of narrow molecular weight fractions, analyzed by light scattering, of the polymer to be studied.

The use of G.P.C. is convenient and quite popular because it is rapid, simple and provides a complete description of the entire MWD curve. However, its application for polyolefins [37, 76] is not free from methodological uncertainties and limitations, connected to the necessity of using high temperature to dissolve and destroy polymer cristallinity, and to a certain effective incapacity of fractionation, which can cause an insufficient resolution particularly for high molecular weight fractions and, therefore, their over — or under — estimation. This is even truer if the polymer, as in high density polyethylene, is not always perfectly linear [38]; since its fractionation also depends on parameters (i.e. structure) other than molecular weight, complementary techniques must be resorted to. Obviously, in comparing polymer average molecular weights and their distributions obtained by different researchers, one must critically evaluate the analytical technique utilized, the thereoretical assumptions upon which the technique is based and its experimental accuracy. In other words a quantitative comparison should be made only with great care.

Approximate evaluations of MWD can also be drawn through empirical correlations with the data obtained by measuring some of the polymer melt properties [30–32, 35, 39] as, for example, a "melt flow rate ratio". However, also in this case, as in all correlation attempts [35, 40–42, 179] involving the behavior of polymeric materials, both in the molten and the solid state or in solution, great care must be taken in drawing conclusions.

2.2 Influence of MWD on Polymer Properties

For high density polyethylene the MWD represents one of the basic parameters which, together with average molecular weight, density, short and long chain branching content, determines the physical, mechanical and rheological properties of the polymer [26–29, 249] and therefore its end uses (Table 1).

Although it is difficult to evaluate the influence of each single property on the whole polymer performance, it can be said in general that, average molecular weight

Table 1. High density polyethylene: specifications and applications of typical commercial products[a]

Type of processing and application	Intrinsic viscosity[b] (dl/g)	Melt flow index[c] (g/10′)	Polydispersity index[d] Q	Density[e] (g/cm³)
Extrusion pipes; blow molding film and large size containers	2.8–3.0	MI$_5$ 0.2–0.4	15–25	0.945–0.955
Blow molding bottles	2.2–2.6	MI 0.2–0.4	10–15	0.948–0.956
Extrusion monofilament and tapes	1.8–2.1	MI 0.5–1.0	7–8	0.948–0.952
Injection molding	1.0–1.6	MI 2–40	3–5	0.950–0.960

[a] Some products are sligthly modified by copolymerization of ethylene with α-olefins;
[b] In THN at 135 °C;
[c] ASTM D 1238;
[d] G.P.C. data;
[e] ASTM D 1505

being the same, the broader the MWD of a polyethylene, the easier is its processability. In other words polyethylene with a broader MWD shows greater flowability in the molten state at a high shear rate (an important property for the blowing and extrusion techniques). On the other hand, items made from polymers with a narrow MWD have greater dimensional stability, higher impact resistance, greater toughness at low temperatures and higher resistance to environmental stress-cracking. Moreover, when considering polyethylene application properties, it is important not only to evaluate the MWD width, but also the shape of its curve [137, 258].

Concerning isotactic polypropylene, whose properties are essentially determined [27] by the amount of crystallinity and the average molecular weight, MWD has relatively less influence on polymer properties. Commercial products, no matter which end use have to fit, show MWD with polydispersity values generally between 6 and 12. Broader MWDs are not required since polypropylene gives practically no surface or distortion defects during extrusion. Narrower MWDs (Q < 5), which are useful, for example, in rapid molding of products with good mechanical properties, are not obtained through polymerization at present, but rather by thermomechanical break-down of the polymer.

3 Origins of MWD

Along with the practical importance of knowing the MWD of a polymer, the theoretical and interpretative aspect also exists. This aspect is no less significant since MWD, as a recording of what happens during polymerization, could give an important contribution in understanding the mechanisms which govern this reaction.

Ziegler-Natta polymerization is characterized by a series of elementary reactions which can be represented by suitable models. A scheme of such reactions, as proposed by Grieveson [56] and including Natta's [81] original hypotheses as well, shows (Table 2) besides the initiation and propagation steps, the various possible types of chain transfer and termination processes, both in the presence and in the absence of

Table 2. Kinetic scheme for Ziegler-Natta polymerization (ethylene)

Initiation:

$Cat^*-R + CH_2 = CH_2 \rightarrow Cat^*-CH_2CH_2R$

or $Cat^*-H + CH_2 = CH_2 \rightarrow Cat^*-CH_2-CH_3$

Propagation:

$Cat^*-(CH_2-CH_2)_n-R + CH_2 = CH_2 \rightarrow Cat^*-(CH_2-CH_2)_{n+1}-R$

Termination:

— spontaneous

   $Cat^*-(CH_2-CH_2)_n-R \rightarrow Cat-(CH_2-CH_2)_n-R$

— by transfer with hydride β-elimination

   $Cat^*-(CH_2-CH_2)_n-R \rightarrow Cat^*-H + CH_2=CH-(CH_2-CH_2)_{n-1}-R$

— by transfer with aluminium alkyl

   $Cat^*-(CH_2-CH_2)_n-R + AlR_3' \rightarrow Cat^*-R' + R_2' Al-(CH_2-CH_2)_n-R$

— by transfer with hydrogen

   $Cat^*-(CH_2-CH_2)_n-R + H_2 \rightarrow Cat^*-H + CH_3-CH_2-(CH_2-CH_2)_{n-1}-R$

— by transfer with monomer

   $Cat^*-(CH_2-CH_2)_n-R + CH_2=CH_2 \rightarrow Cat^*-CH_2-CH_3 + CH_2=CH-(CH_2-CH_2)_{n-1}$

   $-R$

[a] Cat* is an active center; Cat is a deactivated center; R and R' are alkyl groups

some added chain terminating agents. An even more global kinetic scheme was recently proposed by Böhm [178] and by Zakharov et al. [206].

In principle, with the various elementary processes and the relevant kinetic constants, it should be possible to formulate [33] a mathematical expression which links MWD to such constants. Vice versa experimental knowledge of the polymer MWD function should permit to ascertain the polymerization mechanism. Actually, only in particularly simple cases, exemplified by Peebles [43] with regard to homogeneous polymerization processes, the MWD function can be analytically determined, on probabilistic basis, as a simple function of the molecular weight or of the macromolecular chain length. In particular, when the polymerization mechanism involves, as in the Ziegler-Natta catalysis, a macromolecular chain termination step caused by a transfer process or by a spontaneous deactivation of the active center, MWD should assume the characteristics of the "most probable" distribution curve according to Schulz and Flory, with a Q value of about 2 [43-45, 48, 55].

As can be expected, the polydispersity index Q, instead, assumes a value very close to 1 and the shape of the MWD curve comes closer to that of Poisson, in those particular cases free from transfer and termination processes [247].

All this has been experimentally proved (in agreement with the proposed kinetic models though they cannot all be referred to the same mechanism) both in ethylene and propylene [245] *oligomerization* (homophase system) and in some cases of ethylene and propylene *polymerization*, provided homogeneous catalytic systems are used.

For example, Henrici-Olivè and Olivè [47,48] took into consideration the ethylene oligomerization by some soluble catalytic systems which, under definite conditions, almost exclusively yield linear α-olefins. In this case, without secondary reactions (which would bring about branching and insaturation of macromolecular chains) and without chain-termination, a mechanism was hypothized which requires only a chain propagation reaction and a chain transfer reaction with the monomer. According to such a mechanism, a "Schulz-Flory" type MWD was calculated fitting well experimental data, even though obtained by different authors [49-53,] with different catalytic systems. Therefore Henrici-Olivè and Olivè concluded that not only all the catalytic systems follow the same mechanism, but that it also is possible, with such a model, to calculate the optimal weight fraction for every oligomer.

Matkovskii et al. [46] also studied ethylene oligomerization. They found that with the homogeneous $TiCl_4-AlC_2H_5Cl_2$ catalytic system in benzene and at room temperature, the MWD of oligomers was quite similar to the "Schulz-Flory" distribution and its experimental average values, determined by G.P.C., agree with the theoretical ones derived by analyzing a kinetic model in an almost stationary approximation.

A detailed study of the oligomerization of ethylene in toluene with soluble catalytic systems of the type $(C_5H_5)_2TiRCl-AlC_2H_5Cl_2$ has been carried out by Fink et al. [251]. By means of plug flow reactor in totally homogeneous phase, they were able to prove, for the very initial instants of the polyreaction, a good agreement between the experimental distributions and the theoretical ones calculated from a reaction scheme established by kinetical analysis and mathematical modelling.

As for ethylene polymerization, Belov et al. [54] found that, by using the $(C_5H_5)_2TiCl_2-Al(C_2H_5)_2Cl$ catalytic system soluble in toluene, a good agreement was achieved between the experimental MWD data (near the Schulz-Flory "most probable" distribution for long reaction times) obtained by polymer fractionation and the theoretical data calculated on the basis of a kinetic model which assumes (at Al/Ti ratio ≥ 5) a termination by chain transfer reaction with $Al(C_2H_5)_2Cl$ and no chain transfer with the monomer.

Yet Chien [164] demostrated that it would be possible to obtain polyethylene with a Q value near the theoretical 2, with the homogeneous $(C_5H_5)_2TiCl_2-Al(CH_3)_2Cl$ catalytic system, only if carefully controlled *pseudosteady-state* conditions are employed. In fact he showed mathematically that the relatively high experimental polydispersity (Q from 2 to 5 in function of reaction time), is a natural consequence of a polymerization kinetic model [165] based on non stationary first order initiation, chain propagation and bimolecular chain termination by recombination.

However, in reality, polyolefins prepared by using Ziegler-Natta catalytic systems (which mainly are heterogeneous) show [61-78] a polydispersity index which is generally greater than two, reaching Q values from 5 to 10 and even beyond. In the case of polypropylene, Q value can be higher than 10 and in the case of polyethylene even beyond 20.

Some empirical functions of MWD have been proposed in order to adequately describe experimental data. The one proposed by Tung [79] is represented, in its integral form, by:

$$I(M) = 1 - \exp(-aM^b)$$

where M is the molecular weight and a and b are two experimentally determinable parameters.

The distribution proposed by Wesslau [66] instead, is found in the form:

$$I(M) = (1/\beta\pi^{1/2}) \int_0^M (1/M) \exp[-(1/\beta^2)(\ln M/M_0)^2] \, dM$$

and represents a logarithmic normal distribution in which $\ln M_0$ is the average of the normal distribution of $\ln M$ and β is the standard deviation of this distribution multiplied by the square root of two. Calculated values from such empirical functions satisfactorily agree with the experimental fractionation data both for polyethylene [66, 139] and for polypropylene [70, 77, 80]. The same is true for commercial unfractioned polypropylenes [72] whereas, for moderately degraded polypropylene samples, or for those with relatively high molecular weight, a better agreement is found with the modified function proposed by Davis et al. [20].

In order to explain the high Q values found, especially for polymers obtained with heterogeneous catalytic systems, even though operating under *pseudostationary* conditions, further hypotheses considering such heterogeneity must be formulated. It should be also noted that these catalytic systems are generally used under such conditions as to yield crystalline polymers (both polyethylene and polypropylene) which are insoluble in the polymerization medium.

Therefore, one is dealing with a heterophasic reaction which could be controlled by typical kinetic factors such as: a) formation and decay of active centers with time, b) presence of a multiplicity of active centers energetically, structurally and chemically different form one another and therefore having different kinetic constants. Moreover a role could also be played by true physical phenomena such as: a) variety of growing chain lifetime depending on the different degree of active centers encapsulation in the polymeric matrix, and b) limitations to heat transfer and, above all, to mass transfer from the gas phase to the liquid phase, from liquid to polymer surface and from the polymer to the surface or to the interior of the catalyst.

Along with these phenomena, physical-chemical phenomena connected with the macromolecular chain desorption from the catalyst surface may also be important.

Therefore, to arrive at an understanding of the factors which govern MWD and at their practical use in the preparation of polyolefins, the above-mentioned parameters should be discussed in detail in connection with the chemical and diffusion hypotheses on the MWD origin.

3.1 Diffusion Theory

As it is well known, in the most important Ziegler-Natta processes, the polymer grows as a semi-crystalline powder on the solid catalyst and precipitates in the reaction medium.

Therefore, it is easy to understand how some experimental evidences, such as the catalytic activity decay, have been attributed to diffusion limitations [168, 180, 181, 194–196, 199, 200]. Some of these limitations, connected to mass transfer of the monomer from the gas to the liquid phase and from the liquid phase to the polymer surface, can be generally eliminated or at least minimized by means of an appropriate choice of operating conditions [13, 246].

However, much uncertainty remains as to the role played by the monomer diffusion to the active center through the polymer layer which covers it or through the pores of the catalyst itself.

The hindrance to monomer diffusion toward the surface of the catalyst granule caused by its progressive encapsulation by the polymer, was first claimed by Pasquon et al. [180] in order to explain the activity decay with time of the $TiCl_3-Al(C_2H_5)_3$ catalytic system. Many authors have since associated such a concept to the broadening of MWD.

Coover [181] attributed this phenomenon to the slowing down of the chain transfer reaction rate, which would be caused by the increasing limitation to the contact between aluminium alkyl and the active centers, due to their progressive occlusion by polymer.

Buls and Higgins [182], by applying their "Uniform site" theory of Ziegler catalysis to propylene polymerization with $TiCl_3-Al(C_2H_5)_2Cl$, formulated a mechanism which would justify MWD broadening on a diffusion basis [183]. In particular, by considering the polymer particle formed of concentric layers where locally Q is equal to 2 but where monomer concentration during polymerization varies as a function of the particle radius, polydispersity values were found which agree with experimental data, especially when computations of data pertaining to the oligomers obtained (which could easily be eliminated during the polymer isolation phase) were excluded. However, the authors neither justify on an experimental basis, nor quantitatively specify the resistance to the monomer diffusion throughout the polymer matrix to the active center.

Subsequently Crabtree et al. [169] developed a model according to which polymerization would take place with the encapsulation of catalyst sub-particles. Such a model provides for a rapid increase in the degree of polydispersity to a maximum and then a slow drop with increasing polymer yield or reaction time.

Schmeal and Street [186] and Sing and Merril [187] had previously proposed various polymer growth models and, by dealing with them mathematically, however without supplying experimental data, they reached the conclusion that a sufficiently large value of Thiele *modulus* [233], can account for broad MWD. Such parameter can be written in the form:

$$\alpha = \left(\frac{k_p C_0^*}{D_m}\right)^{1/2} S_0$$

Where: k_p is the propagation rate constant, C_0^* the initial concentration of active
centers, D_m the monomer diffusivity, and S_0 the initial radius of the
catalyst particle.

On the contrary, for low values of the Thiele *modulus*, reaction control prevails and low values of MWD (Q = 2) are predicted.

The simplest of the diffusive models proposed, the "Solid core" model, is based on a spherical catalyst particle with a spherical shell of polymer growing around it.

According to Ray et al. [184-185], who recently have reviewed and reproposed the diffusion hypothesis, such a model does not justify the typical polydispersity values of Ziegler-Natta catalysis, since it supplies Q = 2 for realistic physical and kinetic parameters. This is due to the fact that, although according to this model there is a notable limitation affecting the monomer diffusion, the olefin concentration on the catalyst surface keeps patrically constant if the sharp drop, which takes place as polymerization begins, is excluded. On the other hand, the "Solid core" model is certainly poorly indicative of the phenomena which take place during polymerization such as the generally accepted catalyst granule break up and its dispersion in the polymeric matrix [192].

Polymer growth models which are more sophisticated and adhere more closely to reality have recently been reproposed [184–185]. These are the "Polymeric flow" and the "Multigrain" models.

In the "Multigrain" model, fractured catalyst microparticles are produced during the polymerization and uniformly dispersed in the polymer; each of these particles behaves as a *micro* "Solid core" and diffusion within them, as well as in the interstices between them, can take place. In the "Polymeric flow" model the catalyst microparticles are dispersed in a "polymer continuum" and move outward in proportion to the volumetric expansion due to polymerization; only one value of diffusivity is considered. Both these models predict significant MWD broadening due to mass transfer limitations ($Q_{max} \sim 9$ for polypropylene in the "Polymeric flow" model) on the basis of mathematical calculations carried out assuming reasonable values of the kinetic and physical parameters.

Therefore, in principle, MWD broadening might be adequately explained and justified on the basis of diffusion phenomena. However, whether such phenomena agree with other experimental evidences relative to Ziegler-Natta catalysis or not, remains to be proved.

Also the activity decay with time of many catalytic systems may be attributed to their incapsulation by the polymer and to the consequent monomer diffusion hindrance. On the other hand, the activity decay can also be explained on the basis of chemical deactivation of active centres [13].

Other well known phenomena, listed below, connected to Ziegler-Natta polymerization, provide further experimental evidences which could agree with the two last models proposed:

— the extreme *sub-division* of catalyst particles in the polymeric mass [183–188];
— the proportionality of catalytic activity to the total area of the catalyst crystallites ("working surface") [5];
— the phenomenon of catalyst granule shape *replica* by the polymer (even if of much larger size) [193].

Such phenomena give rise to the assumption of the following polymer formation mechanism:

— monomer access to the surface of all *primary* particles which make up the catalyst granule;
— monomer polymerization on such catalyst surface;
— break up of the catalyst granule under the stresses created by solid polymer growth;
— formation of polymeric granules having catalytic microparticles (with dimensions probably of the order of magnitude of crystallites), intimately and uniformely dispersed in it.

Since the individual *primary* particle is impermeable to the monomer and not susceptible to further fragmentation, it should behave as a *micro* "Solid core" causing the formation of polymer *microglobules* independently of its original form [190]. There is no microgranule *replica*, but only *replica* of the granule in its totality, as also stated by the basic theory which explains the *replica* phenomenon.

In this way the "Multigrain" model would take on a precise physical meaning. The microglobular polypropylene surface, obtained with both traditional [239] and "high yield" catalysts [198] is clearly shown out by electron scanning microscopy.

Contrary to the afore mentioned Authors, Chien [191], considering various poly-ethylene and polypropylene catalytic systems, on the basis of the Thiele *modulus* criteria ($Q \geq 5$ for $\alpha > 5$ and $Q \sim 1$ for $\alpha \leq 1$) recently concluded that Ziegler-Natta catalysis is not controlled by diffusion phenomena while Phillips catalysis is.

Such a different conclusion can be understood by considering the difficulties con-nected to the experimental determination and the definition of Thiele *modulus* para-meters, such as S_0 and D_m. According to Chien, S_0 means the catalyst primary particle size with a value of about 10^{-4} cm for α-TiCl$_3$; instead, in the "Multigrain" model, S_0 seems to correspond to the size of the whole catalyst granule.

A source of further confusion is the monomer diffusivity D_m, since it is sometimes associated with the diffusion *inside the catalyst pores* [197] while, in other cases, with diffusion *through polymeric film* which, in turn, can be either amorphous, crystalline or partially crystalline. As a result, D_m can vary even by more than one order of magnitude depending on the type of the above assumptions. Therefore, it seems evident that, although they can adequately explain many experimental phenomena, the physical models suffer from a certain ambiguity.

Furthermore a great deal of experimental evidences exists which does not agree with the diffusion hypothesis.

In the first place, polymers with broad MWD have been found when using hetero-geneous catalytic systems under conditions free from diffusion limitations, such as the case of polymerization carried out under such conditions as to keep the polymer dissolved in the reaction medium [168].

Analogous results have been obtained when operating under such conditions (low conversion, low pressure, etc.) as to guarantee a strictly kinetic control of polymeri-zation [11, 140, 189].

Furthermore, soluble catalytic systems containing homogeneous active centres yield polymers with very narrow MWD ($Q \sim 2$) even when they are used in hetero-phasic processes (see Section 4.1.1).

In the case of heterogeneous "high yield" catalysts, Thiele *modulus* criterion would predict an appreciable broadening of MWD in comparison to traditional catalysts (higher $k_p C_0^*$ values). This broadening could be also due to a notable enlargement of the polymer particle and, therefore, to an increase in the Thiele *modulus* in proportion to catalytic yield [11].

The "high yield" catalysts for polypropylene, which have recently been developed by Montedison in cooperation with Mitsui Petrochemical [146], though showing cristallite sizes similar to those of δ-TiCl$_3$, have activities even higher than two orders of magnitude. However, this does not translate into a substantial broadening of MWD as compared with traditional catalysts. In fact [198], the degree of polymer poly-dispersity depends mainly on the chemical composition of the catalyst and, in some cases, it is actually lower than that obtained with TiCl$_3$—Al(C$_2$H$_5$)$_2$Cl (Table 3).

Even in the case of polyethylene, no dependence has been found [209] of MWD on chemical-physical parameters of the supported catalysts (Table 4).

Another criterion linked to the diffusion hypothesis, suggested by Sing and Merril [187], provides for MWD narrowing through partial poisoning of the catalyst, since it causes a lowering of Thiele *modulus* (lower C_0^* values). It was found, on the contrary that, by treating TiCl$_3$ with various poisons, it was possible to obtain even

Table 3. Polydispersity index of polypropylene[a] samples obtained with different catalytic systems[b]

Catalytic system	$[\eta]$ 135 °C THN (dl/g)	$\bar{M}_n$ $\times 10^{-3}$	$\bar{M}_w$ $\times 10^{-3}$	Q[c]
TiCl₃(AA)-Et₂AlCl	2.05	55	330	6.0
High yield (type A)	2.09	71	337	4.7
High yield (type B)	2.28	58	393	6.8

[a] Boiling heptane insoluble fraction;
[b] Polymerization conditions: hexane slurry at 70 °C and 700 Pa for 4 hours;
[c] G.P.C. data

Table 4. Polydispersity of high density polyethylene obtained with different catalytic systems[a]

Catalytic system	X.R. Crystallinity	Average crystallite size (Å)	B.E.T. Surface area (m²/g)	Catalyst yield (g. Pol/g × Ti × h × atm)	Polydispersity index Q
A	Crystalline	195	25	13000	8.0
B	do	290	10	17000	6.0
C	do	350	75	23000	7.5
D	amorphous	—	<5	4000	4.0
E	do	—	<5	5000	5.0
F	partially crystal.	100	<5	4000	3.5
G	crystalline	85	250	10000	8.0
H	partially crystal.	—	90	1000	12.0

[a] Polymerization conditions: solvent hexane; temperature 75 °C; pressure 1100 Pa; i-Bu₃Al as cocatalyst and hydrogen as molecular weight regulator (Polymer Melt Flow Index 0.1–0.2 g/10′);
[b] G.P.C. data

a considerable broadening of MWD [185, 242]. Similar results were observed by Kashiwa [148] by treating a "high yield" MgCl₂ — supported catalyst for polypropylene with progressive quantities of an electron donor. Although such experimental evidences seem to be in contrast to the diffusion hypothesis they do not exclude the formation of new active centres or modification of those already existing. The centers having different stereospecificities and, among them, different Lewis acidities, were previously described by Pino [250], for similar catalytic systems. The plurality of such sites, is at the basis of chemical theories as to the origin of MWD.

3.2 Chemical Theory

Natta [85] suggested various types of active sites on the surface of solid catalyst as a possible explanation for MWD broadening.

Along with such a purely chemical theory, some authors have proposed models which justify MWD broadening on the basis of chemical-physical phenomena which could take place on the surface of solid catalyst. Mussa [201] associated polymer MWD to the individual lifetimes of the single macromolecules, assuming that these lifetimes are distributed according to the Gauss formula, possibly perturbed by factors limiting the chain growth. The chain propagation rate would increase with an increase in the length of the macromolecule due to an autocatalytic effect produced either by a local overheating or by an increased monomer concentration.

Gordon and Roe [86] paid their attention mainly to chain transfer rather than chain propagation process. Thus, in a hypothetical model the termination rate constant depends on the chain length (due to control of the termination reaction by reversible desorption of the polymer macromolecule from the catalyst surface). A two parameter expression for the MWD function was derived which, with a chain transfer agent, should imply a Q decrease toward an asymptotic value of 2 [202]. On the other hand, according to Roe, no modification of MWD could be expected if catalyst *non-uniformity* were responsible for the broad distribution. More recent data seem to support this latter hypothesis rather than the mechanism of Gordon and Roe.

In fact, MWD obtained at different deuterium-ethylene ratios are not influenced by the amount of the chain terminating agent [203]. Similar results have been found by Berger [7] and Grieveson [56] in ethylene polymerization, using different hydrogen concentrations.

These findings strongly support the hypothesis of polymerization centres *non-uniformity* with regard to the respective rate constants of propagation and termination; each of these centres yields polymers having the "most probable" MWD but different average molecular weights. The total MWD is thus given by *superimposing* the individual "most probable" distributions.

This concept was first mathematically developed by Clark and Bailey [83, 84] according to an exponential distribution of active centres with respect to adsorption energy. The Langmuir-Hinshelwood mechanism for adsorption and reaction was found to fit experimental results (ethylene polymerization over chromium oxide-silica-alumina catalyst) more closely than the Rideal mechanism.

The hypothesis regarding the heterogeneity of active centres as an origin of MWD broadening has been recently reproposed by Vizen and Yakobson [189] for the polymerization of polypropylene at 30 °C with the δ-TiCl$_3$—Al(C$_2$H$_5$)$_2$Cl catalytic system. They examined the MWD of the isotactic part of the polypropylene obtained, under kinetic control, during living polymerization with slow initiation and chain transfer to the monomer according to the following mechanism:

$$n^* + M \xrightarrow{k_i} M_1^* \qquad \text{initation}$$

$$\left.\begin{array}{l} M_1^* + M \xrightarrow{k_p} M_2^* \\[4pt] M_i^* + M \xrightarrow{k_p} M_{i+1}^* \end{array}\right\} \quad \text{propagation}$$

$$M_i^* + M \xrightarrow{k_t} P_i + M_1^* \quad \text{chain transfer}$$

By solving a differential equation system which describes the above polymerization model for homogeneous active centres (k_p = const.), a polydispersity Q value between 1 and 2 was obtained, which cannot explain the MWD of the experimentally obtained polymers (Q = 2.5–5.0). On the other hand, even an exponential distribution of active centres of the form:

$$\delta = \frac{n_0^* C}{e^{-Ck_p^{min}} - e^{-Ck_p^{max}}} \, e^{-Ck_p} \, dk_p$$

(where C is a parameter which characterizes the width of the exponential distribution, k_p^{min} and k_p^{max} are the propagation rate constants for the least and the most active centres, n_0^* is the number of active centres relating to the stationary polymerization rate) leads to a numerical function of MWD which is similar to the Shulz-Flory "most probable" distribution. In order to explain the fairly broad distribution observed, a set of several functions, with different C values for each of the active centres, has been considered. Estimation has shown that four is the lowest possible number of these functions in order to have satisfactory agreement with experimental results. With hydrogen as chain transfer, G.P.C. curves fit well the calculated values. These results indicate that the proposed heterogeneity of centers, according to the propagation rate constant, permits a satisfactory description of the form of distribution curves and of their dependence on polymerization time and hydrogen concentration.

However, the distribution of the activities of centres should be proved with more direct experimental evidences and eventually explained in terms of physical and chemical surface characteristics. Otherwise, according to Gordon and Roe, the proposed distribution of centres is only a mathematical formality to explain almost any MWD a posteriori.

In reality various direct experimental evidences exist to support such an hypothesis first of which, in the case of polypropylene, is the well known heterogeneity according to stereospecific properties of centres [82, 204]. Actually, the stereoregular polypropylene isotactic fraction and the amorphous one generally show broad MWD too, thus proving the probable existence of different active centres although similar in stereoregulating ability [172, 175].

Recently Rishina and Vizen [176] have studied propylene polymerization using δ-TiCl$_3$—Al(C$_2$H$_5$)$_3$ and δ-TiCl$_3$—Al(C$_2$H$_5$)$_2$Cl in the presence of CS$_2$, a powerful catalyst poison. The selective poisoning of polymerization centres showed that all of them, both stereospecific and nonstereospecific, are characterized by a range of values of propagation rate constants. Pino and Mülhaupt [250], investigated the stereoregularity and microstructure of polypropylene obtained with "high yield" MgCl$_2$-supported catalyst at various Lewis base-AlR$_3$ ratios; they were able to show that both the stereospecific and nonsterospecific centres differ in their Lewis acidity.

On the other hand, the presence of many types of active species seems to be a logical consequence of the chemical-physical structure of the solid catalytic component.

Both traditional and supported heterogeneous Ziegler-Natta catalysts consist of aggregates which, during polymerization, are completely disrupted into the single crystallites. The crystalline modifications of both the α and β forms of TiCl$_3$ have been described by the Natta School as consisting of layers of titanium ions sandwiched between chlorine ions and differing in the shape of the unit cell; the δ modification exhibits a disordered structure resulting from translation and rotation of the chlorine planes.

The structures of the crystalline forms of activated $MgCl_2$, a very effective support for "high yield" catalysts preparation, are very similar to those of $TiCl_3$ [227]. In order to maintain the electroneutrality of the system, these crystalline structures produce different surface sites having chloride ions vacancy.

This concept led Cossee and Arlman [232] to propose for $TiCl_3$ a believable model for the active centre. Recently this hypothesis has been extended, with convincing results, to the examination of a particular type of $TiCl_3$ [228], while Corradini et al. [216, 217] have developed an analysis for models of Ziegler-Natta stereospecific polymerization on the basis of non-bonded interactions at the catalytic sites on the surface of layered modifications of $TiCl_3$. This model has been recently extended to the $MgCl_2$-supported $TiCl_4$ catalyst [252].

To conclude, the hypothesis that different types of active polymerization centres are present seems to be reasonable. The following well known experimental observations can further testify in favour of the theory based on active centre plurality:

— perfectly soluble catalytic systems containing homogeneous active centres make it possible to obtain narrow MWD even in the case of heterophasic (slurry) polymerization;
— it is possible to change polymer MWD by modifying the chemical composition of the catalyst;
— it is possible to obtain broad MWD even when operating under such conditions as to have both the catalyst and the polymer dissolved in the reaction medium. In this case the presence of various active species has been actually demonstrated, by suitable analytical techniques.

The surface heterogeneity has been recently supported by Keii et al. [253], in detailed kinetic and fractional studies of polypropylene obtained with both conventional and $MgCl_2$-supported "high yield" catalytic systems.

Analogously with MWD broadening, even the composition heterogeneity observed for ethylene-propylene copolymers, obtained with either homogeneous or heterogeneous catalytic systems, has often been attributed to the plurality of active centres [98, 229–231, 234, 244].

Moreover the narrower MWD of some ethylene-propylene-diene terpolymers, in comparison to that of ethylene-propylene copolymers, has been ascribed [241] to the preferential deactivation of some active centers.

4 Control of MWD

MWD control in Ziegler-Natta polymerization of olefins will be dealt with in this chapter, by sub-dividing the topic into the two following general headings:
4.1 Catalytic system;
4.2 Polymerization parameters.
Finally some considerations on industrial polymerization processes and patents will be given.

In examining this matter, the comparison of the various data from the literature requires caution because of both the uncertainties inherent in the measurement of polymer polydispersity, and the different polymerization conditions which have been adopted and which often have not been clearly indicated.

4.1 Catalytic System

One can, in principle, assume that at least the following characteristics might contribute to the behaviour of the catalytic system: physical state (homogeneous or heterogeneous); type of transition metal, its oxidation state and its electronic environment, thus including type, number and geometry of the ligands; type of cocatalyst. In the case of heterogeneous catalytic systems one must also find out whether the transition metal compound constitutes the crystalline lattice of the catalyst or whether it is supported on a carrier. Concerning catalytic systems for polypropylene, also the presence of a Lewis base, as well as its nature, has to be considered, along with the above-mentioned factors.

4.1.1 Physical State

Homogeneous catalytic systems can often be chemically and structurally better defined than the heterogeneous ones and are free from physical surface phenomena. Thus they are more suitable for the study of Ziegler-Natta catalysis mechanisms in general, and of the polymer MWD in particular. Furthermore, due to their nature, they should produce polymers with particularly narrow MWD. This, however, does not preclude the possibility of varying MWD by changing types and concentrations of individual active species in these catalytic systems. Unfortunately they have poor industrial interest as they usually show low activity and give rise to polymers with bad powder morphology and no (or low) stereoregularity in the case of isotactic polypropylene. Important exception are those vanadium-based soluble systems (or which are soluble at least in the initial moments of polymerization) for ethylene-propylene elastomer production [240].

An excellent review on soluble systems, including also the more recent progresses in ethylene polymerization, is that of Sinn and Kaminsky [16], who even developed soluble catalytic systems with extremely high activities by using alumoxane and biscyclopentadienyl-titanium or-zirconium compounds. Interesting results have also been obtained in ethylene oligomerization [94], and in propylene dimerization [89].

Examples of homogeneous catalytic systems in propylene polymerization are very few. Zambelli et al. [87,88] obtained syndiotactic polypropylene by working at low temperatures with catalytic systems based on some vanadium compounds and aluminum alkyls, while Giannini et al. [90] were the first to prepare isotactic polypropylene with some benzyl derivatives of titanium or zirconium.

Unfortunately, despite the great number of papers and the doubtless interest in both kinetic and mechanistic studies on the behavior of the homogeneous catalytic systems, many aspects of this part of Ziegler-Natta catalysis remain in the dark. To this point reference is made to the works and reviews of Henrici-Olivé and Olivé [91,92], Ballard [93], Reichert [57,58] and finally of Fink et al. [59,60,251] on the MWD study of ethylene oligomers. More recently Reichert [23] summarized the status of $(C_5H_5)_2-Ti^{IV}$-type soluble catalytic systems in combination with various aluminum alkyls.

It is difficult from what is known about homogeneous catalysis to infer something which can be applied directly to heterogeneous catalytic systems which are characterized by a low concentration and a plurality of active species on the catalyst surface; furthermore the determination of active centers is still debatable [2,5,6]. For these reasons the influence of some chemical factors, such as the transition metal ligand, has been studied with soluble catalytic systems which, often, have a better defined chemical composition and, sometimes, also a more clear molecular structure and are certainly free from complications arising from the heterogeneous nature of the surface.

That homogeneous catalytic systems contain only one or very few active species has been confirmed by the fact that they generally yield polymers with a MWD varing from narrow to very narrow, with Q values approaching the theoretical value of 2. The results obtained for the ethylene polymerization by Breslow and Newburg [95], by Belov et al. [54] with the $(C_5H_5)_2 TiCl_2—Al(C_2H_5)_2Cl$ catalytic system, and by Carrick [96] with the $VCl_4—Sn(C_6H_5)_4—AlBr_3$ system point in this direction. On the other hand, with the $(C_5H_5)_2TiCl_2—Al(CH_3)_2Cl$ catalytic system, Chien [164] found $Q \geq 2$ according to polymerization conditions, but he was able to justify such a result on the basis of a kinetic mechanism which implies a non — stationary initiation stage. With $(C_5H_5)_2TiCl_2—Al(C_2H_5)_2Cl$ (at Al/Ti ratio 2.5) Höcker and Saeki [107] have studied products deriving from the polymerization of ethylene, for short times, in benzene at 5 °C. From the I. R. and N. M. R. spectra of the polymers they concluded that the low molecular weights obtained consisted of saturated hydrocarbon chains, thus indicating the absence of termination or transfer reaction. Accordingly the MWD ($Q = 1.07$) was very close to the Poisson distribution. Furthermore, as will be seen later, a broadening of this MWD would derive from electronic variations undergone by the titanium-carbon bond, introducing appropriate substituents in the cyclopentadienyl ring.

Instead, in the oligomers prepared with $TiCl_4—AlC_2H_5Cl_2$ system in chlorobenzene at 0 °C and an Al/Ti ratio of 6, a greater structural complexity was seen due to transfer reactions which bring about [107] an intermediate MWD between that of Poisson and the "most probable" one.

In the attempt to simplify the study of catalytic systems based on titanium cyclopentadienyl compounds, avoiding interference by titanium alkylation and reduction by the aluminum alkyl, Waters and Mortimer [106] studied ethylene polymerization at 0 °C in toluene, with the $(C_5H_5)_2TiRCl—AlR'Cl_2$ soluble catalytic system. For $R = R' = $ ethyl and Al/Ti ratios from 1 to 10, they found $Q = 1.81 \pm 0.38$ from G.P.C. data, independently of any polymerization parameter. Together with the observations that all the titanium-carbon bonds are active for ethylene insertion right from the beginning of the polymerization, and that approximately one macromolecule is found for every titanium atom, this soluble catalytic system gives evidence that most active centres remain active during the whole polymerization; this is considered similar to a living one.

The concept that a homogeneous catalytic system, when formed of a single active species, yield polymers with a narrow MWD has often been cited; viceversa, a narrow MWD is taken as an indirect proof of a polymerization occurring on a single active site.

Thus, results obtained with some vanadium based homogeneous catalytic systems have also been explained in this manner. Christman [97] has examined polyethylene obtained in a completely homogeneous reaction system, since the catalytic system was soluble and the polymerization took place under such temperature conditions as to keep the polymer dissolved in the reaction medium.

With the $VO(O \text{ ter-}C_4H_9)_3—Al_2(C_2H_5)_3Cl_3$ catalytic system and with ethyl trichloroacetate acting as a "promoter", polyethylene with very narrow MWD was obtained ($Q = 2.1$ to 2.6; values from fractionation data). Similar behaviour was observed [97] for a modified "Carrick-like" catalytic system prepared from a soluble vanadium compound, $Sn(C_6H_5)_4$ and $AlBr_3$, in the presence of methyl trichloroacetate as "promoter".

Cozewith and Ver Strate [98] examined fractionation data of ethylene-propylene copolymers obtained with homogeneous or apparently homogeneous systems, based on vanadium compounds such as VCl_4 and aluminum trialkyl, or $VOCl_3$ and $Al(C_2H_5)_2Cl$. While some catalytic systems gave the expected narrow MWD with $Q \sim 2$ and high composition uniformity concerning monomeric units distribution, others gave a wider composition distribution and multimodal and broader MWD with Q even > 10. The authors attributed this last result to different active centres.

From an interpretative point of view it is, therefore, of great importance to determine which and how many active species are in any given homogenous catalytic system, if it is really homogenous and if it remains so throughout the whole polymerization. Unfortunately, visual inspection does not always give certain results.

For example, Soga et al. [99] found, by examining propylene polymerization in toluene in the temperature range 0–65 °C with an apparently soluble catalyst such as tetrabenzylzirconium, that the isotactic index and the polydispersity of the polymer increased as the polymerization temperature increased. Furthermore, the value of $Q > 30$ was much greater than that predicted for homogeneous active centres. From these results the authors concluded that the catalyst should be made up of small invisible colloid-type particles. The same interpretation is valid for the surprising results ($Q \sim 40$) obtained [100] for polyethylene, in Isopar solution at 200 °C, with the originally soluble tetrabenzylzirconium-water catalyst.

Once greater certainty as to the total real solubility of a catalytic system has been gained, a kinetic study can supply information about the cause determining the type of the observed MWD.

A catalytic system which has been greatly studied is based on titanium cyclopentadienyl derivatives. Belov et al. [101–104], studying ethylene polymerization with the $(C_5H_5)_2TiCl_2—Al(C_2H_5)_2Cl$ system, found that while in a non-polar solvent, such as benzene, a polymer with a unimodal MWD is formed having a low polydispersity ($Q = 1.8$), in a chlorinated solvent, such as ethyl chloride, the MWD curve is bimodal and broader. This should be connected to the different kinetic behavior of the catalytic complex which reveals a constant polymerization activity for a long time in ethyl chloride, as opposed to benzene. In order to justify the bimodality of the MWD curve, although the catalytic system was certainly soluble, the authors demonstrated the existence of two types of active centres slightly differing in propagation and termination constants. In fact the original catalytic system is partially transformed into the "blue complex" $(C_5H_5)_2TiCl \cdot AlC_2H_5Cl_2$ which, with ethyl cloride and $Al(C_2H_5)_2Cl$, generates a new type of active centre able to produce, independently from the original one, polymer with narrow and unimodal MWD.

The two types of active centres, were confirmed in that the first peak, which is found in the zone of low molecular weights in the MWD curve, for polyethylene prepared with $(C_5H_5)_2TiCl_2—Al(C_2H_5)_2Cl$, pratically coincides with the position of the single peak obtained with the "blue complex"-$Al(C_2H_5)_2Cl$ catalytic system. A similar behaviour was observed [105] for the $(C_5H_5)_2TiCH_3Cl-AlCH_3Cl_2$ soluble catalytic system with which a variable and very broad (Q from 2 to 50) polyethylene MWD was found by G.P.C. It was also found that the shape corresponding to the broad MWD could vary from monomodal to trimodal. However it was not possible to correlate such behaviour with any characteristic of the system examined. Accidental impurities, such as traces of water, could be the cause of such a behaviour as demonstrated for the $(C_5H_5)_2TiRCl$-alumoxane catalytic system [161-163].

Another widely studied area of homogeneous catalytic systems is based on vanadium compounds which, under specific conditions, polymerize propylene into a syndiotactic polymer.

By using G.P.C. Doi et al. [108] studied a polymer produced at −78 °C with VCl_4 and $Al(C_2H_5)_2Cl$ and observed a unimodal and almost symmetrical MWD curve (Q = 1.8 ± 0.2). Confirming one type of homogeneous active centre, the Q value proved to be independent of polymerization temperature down to −40 °C, while, at temperatures higher than −21 °C, the MWD curve broadened, became bimodal at −10 °C, and finally unimodal again at 41 °C but with Q = 19. This phenomenon was easily interpreted after observing that at low temperatures the homogeneous catalytic active species gives a syndiotactic polymer while, at higher temperatures, a heterogeneous active species gradually predominates, which is responsible for the formation of isotactic polymers.

Keii [109] extended this study finding Q value of 2 ± 0.5 for syndiotactic polymer obtained at −78 °C with the $VCl_4-Al(C_2H_5)_2X$, (X = C_2H_5, Cl, Br) homogeneous catalytic system having trivalent vanadium centres. Instead a Q value of 17 to 19 was found for isotactic polymer obtained with heterogeneous catalytic systems having bivalent vanadium centres (at −78 °C for X = H, I; at 41 °C for X = C_2H_5, Cl).

A similar behaviour was also exhibited by a catalytic system based on a titanium compound which, as opposed to those based on vanadium, yields isotactic polypropylene. In fact, Doi et al. [131] found that the MWD of polypropylene, prepared with the $Ti(OC_4H_9)_4-Al_2(C_2H_5)_3Cl_3$ soluble catalytic system at temperatures below 21 °C, was narrow (Q ≤ 2.0) and monomodal, while the polypropylene prepared at temperatures above 31 °C was enriched by high molecular weight fractions; above 41 °C a very broad and bimodal MWD curve was obtained. The authors were able to demonstrate that a single homogeneous active centre (narrow MWD) was at work below 21 °C, while at higher temperatures, along with this centre, another insoluble heterogeneous centre was present. Confirming two different active centres which act independently, it was found that the first type (homogeneous) produces a polymer with a lower degree of isotacticity than the second one (heterogeneous). Unusually narrow (Q = 1.05–1.25) MWD was found [110,111,235] with the V $(acac)_3-AlR_2Cl$ (R=C_2H_5; i-C_4H_9) system used for low temperature (from −65 °C to −78 °C) propylene polymerization giving a syndiotactic polymer. Such a result, which is quite close to that of a Poisson type distribution, can be experimentally justified by the polymerization, in absence of transfer and termination reactions, taking on the characteristics of a living one.

Therefore, together all the results summarized in Table 5, clearly demonstrate that, in reality, with homogeneous catalytic systems, polymers with narrow MWD can be prepared provided that polymerization conditions are properly choosen and controlled. In fact, polymers with broad MWD and even with bimodal curves, result

Table 5. Influence of catalytic systems on MWD

Catalytic system Type	nature	Monomer	Polymerization conditions	Polymer	Type of polydispersity and Q index	Ref.
$(C_5H_5)_2TiCl_2$-$Al(C_2H_5)_2Cl$	a	ethylene	20 °C, toluene	polyethylene	≦2.0	54)
$(C_5H_5)_2TiCl_2$-$Al(C_2H_5)_2Cl$	a	ethylene	30 °C, toluene or heptane	polyethylene	3.6	95)
$(C_5H_5)_2TiCl_2$-$Al(C_2H_5)_2Cl$	a	ethylene	5 °C, benzene	polyethylene	1.07	107)
$(C_5H_5)_2TiCl_2$-$Al(C_2H_5)_2Cl$	a	ethylene	20 °C, benzene	polyethylene	1.8	101–104)
$(C_5H_5)_2TiCl_2$-$Al(C_2H_5)_2Cl$	b	ethylene	20 °C, ethyl chloride	polyethylene	broad and bimodal MWD	101–104)
$(C_5H_5)_2TiCl_2$-$Al(CH_3)_2Cl$	a	ethylene	30 °C, toluene	polyethylene	about 2 to 5	164)
$(C_5H_5)_2TiRCl$-alumoxane	b	ethylene	0 °C, toluene	polyethylene	from narrow to broad and bimodal MWD	161–163)
$(C_5H_5)_2TiCH_3Cl$-$AlCH_3Cl_2$	b	ethylene	0 °C, toluene	polyethylene	2 or more; monomodal to trimodal MWD	105)
$(C_5H_5)_2TiC_2H_5Cl$-$AlC_2H_5Cl_2$	a	ethylene	0 °C, toluene; Al/Ti = 1–10	polyethylene	1.81 ± 0.38	106)
VCl_4-$Sn(C_6H_5)_4$-$AlBr_3$	a	ethylene	65 °C, cyclohexane	polyethylene	1.5	96)
$VO(O\ ter$-$C_4H_9)_3$-$Al_2(C_2H_5)_3Cl_3$-methyltrichloroacetate	a	ethylene	120 °C, hydrocarbon	polyethylene	2.1–2.6	97)
VCl_4-AlR_3 or $VOCl_3$-$Al(C_2H_5)_2Cl$	b	ethylene and propylene	26 °C, hydrocarbon	ethylene-propylene copolymer	2 to 10 or more	98)
VCl_4-$Al(C_2H_5)_2Cl$	a	propylene	−78° to −40 °C, heptane	syndiotactic polypropylene	1.8 ± 0.2	108)
VCl_4-$Al(C_2H_5)_2Cl$	c	propylene	−21° to +41 °C, heptane	isotactic polypropylene	broad	108)
VCl_4-$Al(C_2H_5)_2X$, (X = C_2H_5, Cl, Br)	a	propylene	−78 °C, heptane	syndiotactic polypropylene	2 ± 0.5	109)
VCl_4-$Al(C_2H_5)_2X$, (X = H, I, Cl, C_2H_5)	c	propylene	−78 °C to +41 °C, heptane	isotactic polypropylene	16 to 19	109)
$V(acac)_3$-$Al(C_2H_5)_2Cl$	a	propylene	−78 °C, toluene	syndiotactic polypropylene	1.05–1.20	110–111)
$Ti(OC_4H_9)_4$-$Al_2(C_2H_5)_3Cl_3$	a	propylene	below 21 °C, heptane	isotactic polypropylene	≦2.0	131)
$Ti(OC_4H_9)_4$-$Al_2(C_2H_5)_3Cl_3$	c	propylene	above 41 °C, heptane	polypropylene	broad and bimodal MWD	131)

a) Homogeneous; b) Homogeneous with more types of active centers; c) Heterogeneous from physical (i.e. heterogeneization) or chemical (i.e. formation of new active species) transformations undergone by the catalytic system just before or during the polymerization; this has often been shown by reliable experimental evidences. The same principle, which brings about a deviation of the polymer MWD curve from the theoretical behaviour predicted, was also assumed in order to justify the composition heterogeneity observed in some ethylene-propylene copolymers [98, 231].

Heterogeneous catalytic systems give broad MWD, as a natural consequence of the plurality of surface active species. However the polymer polydispersity can be controlled, to a certain extent, as will be shown in the next sections.

4.1.2 Transition Metal

As well known, the type of transition metal can deeply affect many fundamental aspects of the Ziegler-Natta catalysis (e.g. activity, stereochemical and molecular weight control). Thus a certain influence on polymer polydispersity could be expected.

In the case of polypropylene, Davis and Tobias [70] observed more than double polydispersity Q index by using $TiCl_3$, instead of VCl_3, together with triethylaluminum as cocatalyst.

Herrmann and Streck [126] recently reported polyethylene and polypropylene MWD variations by varying the type of metal M_v in catalytic systems such as $(TiM_vAl_xCl_yO_z)$ $-AlR_3$ obtained from the reaction of $Al(C_2H_5)_nCl_{3-n}$ with bimetallic μ-oxoalkoxides.

In the patent literature there are numerous indications to obtain polymers with broad MWD using catalysts containing different active centres. These catalysts are prepared by combining compounds of two or more transition metals (Ti, V, Zr, Cr, etc.). This supports the hypothesis that such centres give rise to independent polymerization processes characterized by different rate constants of the elementary stages of the propagation and transfer reactions.

4.1.3 Ligands and Oxidation State of the Transition Metal

The nature of the ligands of the transition metal plays an important role in defining its oxidation state in the active centre. Furthermore, the ligands directly influence the reactivity of the transition metal towards the metal alkyl, the stability of the active centre and its olefin coordination and insertion ability. Thus, by properly varying the steric and electronic nature of the ligands, it should even be possible to alter the kinetic behavior of the active centre.

Novakowska et al. [121] studied the possibility to regulate polyethylene molecular weight by varying the type and number of ligands in $Ti(OR)_nCl_{4-n}-Al_2(C_2H_5)_3Cl_3$. Observations about the influence of the ligand nature both on polymerization rate and on polyethylene molecular weight have also been made by Sangalov et al. [123] and Tajima et al. [124], but MWD was not considered.

The importance of the catalyst chemical composition in MWD regulation was first pointed out by Wesslau [65]. By examining the polydispersity of polyethylene obtained with $TiCl_3-TiX_4-AlRX_2$, he found a narrowing of MWD (Q from 13.2 to 3.4), as a function of the progressive substitution of X groups (X = chlorine atom) with aliphatic or aromatic alkoxide groups; he thus refuted the previous hypothesis [116] that attributed broad polyolefins MWD only to the catalyst physical heterogeneity. Similar results were also obtained by other authors. For example

Schreyer [125] observed that the polyethylene obtained with $TiCl_n(Oi-Pr)_{4-n}-LiAl-(Decyl)_4$, had a narrower MWD when n was 1, 2, 3 instead of 4.

Mortimer [117] obtained a narrowing of polyethylene MWD by adding a proper amount of aluminium alkoxide to a $VOCl_3-AlR_3$ catalytic system.

Höecker and Saeki [107], by G.P.C. analysis of the polyethylene obtained with the $(C_5H_4R)_2TiCl_2-Al(C_2H_5)_2Cl$ soluble catalytic system, were able to observe small variations in MWD according to the electronic influence of $R(R=H, CH_3, C_2H_5)$ on the Ti—C bond. The use of strong electronacceptor ligands of the titanium atom, such as Cl in place of C_5H_5 (as in $TiCl_4-AlC_2H_5Cl_2$), promotes a more frequent β-hydrogen elimination reaction, thus producing oligomers instead of high polymers and also broadening the MWD.

Schindler [118,119] and Henrici-Olivé and Olivé [94,120] justified the influence on molecular weight by the oxidation state of the transition metal and by the electron donor power of its ligands respectively. A decrease in the positive charge on the transition metal, both because it is in a lower valence state (i.e. Ti^{III} compared to Ti^{IV}), and due to the action of an electron donor ligand, decreases the frequency of β-hydrogen transfer reaction in comparison to that of the olefin insertion. While Henrici-Olivé and Olivé correlated ethylene oligomer or polymer formation to the variation in the value of n in the $Ti(OC_2H_5)_nCl_{4-n}-AlC_2H_5Cl_2$ ($n = 0-4$, at Al/Ti ratio 5) soluble catalytic system, Schindler studied the polymerization of the same monomer with $TiCl_4-Al(C_2H_5)_3$ and $TiCl_4-Al(C_2H_5)_2Cl$. In this work he used deuterium, to differentiate the polymers formed by Ti^{III} and Ti^{IV}, and to measure the relative values of $\bar{M}_n$ and $\bar{M}_w$. In this way, it was possible to demonstrate that the polymer formed by Ti^{III} active centres has a much lower index of polydispersity than that formed by the Ti^{IV} sites.

Similar hypothesis has been pointed out [122] to justify that $CH_3TiCl_2-CH_3TiCl_3$, with titanium in two oxidation states, produces, in CH_2Cl_2 at $-70\ °C$, a polyethylene with a broader MWD than that obtained with $CH_3TiCl_2-TiCl_3$ with the transition metal only in oxidation state three.

Similar observations were made by Petrova [136] comparing the MWD of the polyethylene obtained with a high yield catalytic system, such as $(C_6H_5)_2Mg-TiCl_4-Al(C_2H_5)_2Cl$, to that of the polymer obtained with $TiCl_4-Al(C_2H_5)_2Cl$. The Melt Flow Index being equal, the MWD in the first case is narrower ($Q = 4-6$) with lower content of the low molecular weight fraction and absence of the high molecular weight fraction. The first characteristic should be due to the almost total absence of Ti^{IV} sites (at least for high Mg/Ti ratios), while the second would be caused by a greater homogeneity of Ti^{III} sites.

Accordingly, a broadening of MWD should be possible only by creating inhomogeneity in the active centre valences so as to promote different capabilities of monomer coordination and insertion and, thus, different propagation constants. The correspondence between narrow MWD and a unique oxidation state of the transition metal has been also pointed out by Christman [97] for the ethylene polymerization with vanadium compounds-aluminum alkyls homogeneous systems. In this case, addition of a "promoter" causes re-oxidation of the deactivated sites (V^{II}) to the same identical initial ones (V^{III}).

For the polyethylene MWD regulation [137], the chemical composition of a $MgCl_2$-supported-type catalytic system is important, both concerning the transition metal electronic environment and the cocatalyst nature. In fact, electron acceptor ligands, which increase acidity of the transition metal, should be capable of broadening MWD while the opposite effect should be induced by electron donor ligands. Through polyethylene prepared with this catalytic system has a MWD similar to that achievable with traditional catalytic systems, more careful examination of the distribution curve has shown some important differences.

Although they cannot be generalized for every catalytic system, the observations on the influence of the ligands and oxidation state of the transition metal can be verified in some simple cases. For example, with some $MgCl_2$ supported "high yield" systems based on TiX_4 (X = NR_2, OR, Cl) and $Al(i-C_4H_9)_3$ as cocatalyst, it was found [209] that the width of polyethylene MWD is increased with an increase in the electron acceptor power of the ligand, according to the series:

$$Ti(NEt_2)_4 < Ti(Ot-Bu)_4 < Ti(On-Bu)_4 \leqq Ti(OEt)_4 < Ti(OPh)_4$$
$$< TiCl_4$$

Moreover MWD is broadened also with an increase in the number of electron acceptor ligands, for example varying x from 0 to 4 by using $TiCl_x(On-Bu)_{4-x}$ as transition metal compound [209].

4.1.4 Cocatalyst

It has previously been shown how the type of transition metal, its oxidation state, its ligands and, more generally, its chemical-physical environment can determine the polyolefin MWD. However, also the metal alkyl (cocatalyst) plays a role in polymerization centre formation and can, therefore, influence its performance. Nonetheless, it must be pointed out that, since the exact composition and structure of the active species are not known, it is difficult to distinguish between the effect of the cocatalyst and of the catalyst on the performance of the said species.

Though there are no theoretical impediments [215-218] in predicting olefin insertion simply into the transition metal-carbon bond, according to Cossee's well known monometallic model [222-224], and there are many examples [157] of "metal alkyl free" catalysts for olefin polymerization, a great deal of evidences also justifies metal alkyl assistance in the formation and life of the active centre [175,219-221].

This seems valid both for homogeneous and heterogeneous catalytic systems, remembering that, in the last case, the crystalline lattice could be sufficient to stabilize the active centre.

In order to regulate polymer MWD, besides the fact that the metal alkyl action mechanism could be different with homogenous or heterogeneous catalytic systems, one must keep in mind the following main functions performed [23,57-60, 178,206-208] by the cocatalyst (which is generally an aluminum alkyl compound): alkylation and, possibly, reduction of the transition metal; dynamic exchange reactions between transition metal and aluminum ligands; chain transfer with the growing chain; reversible adsorption on the active centre making a species temporarily inactive and finally scavenger activity for the impurities present in the reaction medium. Such functions are influenced by the type of metal, by the nature of its ligands (size, electronegativity, etc.), by the metal alkyl concentration and its stoichiometric ratio, as well as by the contact conditions, to the transition metal compound.

Doi et al. [111] studied the effect of different aluminium alkyls on the polydispersity of syndiotactic polypropylene obtained with the $V(acac)_3$-alkyl aluminium halide soluble catalytic system, at temperatures below $-65\ °C$. These authors found that, by varying the type of aluminum alkyl not only the propagation and transfer rates are changed, but also the polymer polydispersity index decreases in the following order:

$$Al(C_2H_5)Cl_2 \geq Al_2(C_2H_5)_3Cl_3 > Al(C_2H_5)_2Br \geq Al(C_2H_5)_2Cl = 1.05–1.20$$

The phenomena observed were attributed to the different electron acceptor power of the aluminum compounds in agreement with the hypotheses of Henrici-Olivé and Olivé [94]. It was also observed [175], examining the isotactic fraction of polypropylene prepared with δ-$TiCl_3$ and various aluminum alkyls, that MWD width depends on the nature of the ligand and specifically that it decreases in this order:

$$Al(C_2H_5)_2Br > Al(C_2H_5)_2Cl > Al(C_2H_5)_3$$

Wesslau [65] was the first to demonstrate that, along with certain heterogeneous catalytic systems for polyethylene, there is a certain MWD dependence on the type and number of ligands altogether distributed between catalyst and cocatalyst. With $TiCl_4$-aluminium alkyls it was concluded [158] that, by varying the type of aluminum alkyl, polymers are obtained with a very similar MWD curve, but with the maximum shifted according to the nature of the alkyl group. Instead, different results were found by Russian researchers [159–160].

There are few data in the literature as to the influence of aluminum alkyl cocatalyst on polymer MWD obtained with "high yield" catalytic systems.

With the $TiCl_4$-MgO catalyst [159] polyethylene polydispersity was not influenced by either the type or concentration of the aluminum. alkyl, while the use of an aluminum alkyl with a higher reducing power, in the case of some $MgCl_2$-supported catalysts, tended to make active centres more homogeneous and to give a narrower MWD in the following order [137]:

$$AlR_2H > AlR_3 > AlR_2OR, AlR_2Cl$$

It was found [209] that, by polymerizing ethylene at $75\ °C$ with a co-milled catalyst based on $TiCl_4$-$MgCl_2$, the MWD measured by G.P.C., broadens with an increase in alkyl chain lenght of the $Al(C_nH_{2n+1})_3$ aluminum compound (Q from 6 to 12 for n from 2 to 8) due to the preferential increase in the weight average molecular weight. However, it was not possible to find a clear correlation between MWD width and the Lewis acid strength of AlR_2X by varying X nature (X = R, H, Cl, OR).

On the other hand, changes in MWD are expected when the aluminum alkyl undergoes a true transformation following a chemical reaction with a third additional component. Soluble catalytic systems based on titanocenes coupled with alumoxanes proved to be suitable for such study. Cihlář et al. [161–163] investigated the MWD of polyethylene obtained with $(C_5H_5)_2TiRCl$-$(AlR'Cl_2 + H_2O)$; for an Al/Ti molar ratio of 2 (R = R' = C_2H_5), by employing small quantities of water in the system the MWD was broad at molar ratio $Al/H_2O = 0.1$, became bimodal at a ratio of 0.2 and became again monomodal and relatively narrow for values up to 0.67. Once the quantity of water was estabilshed, the MWD was practically independent of both the Al/Ti ratio and the polymerization time. The titanocene with R = CH_3, also gives a narrow MWD while, when R' = CH_3, the distribution curves for any type of R broadened due, in particular, to an increase in low molecular weights. Finally, if the alumoxane is derived from a dialkyl or trialkyl aluminum, polymers with bimodal MWD curves are obtained. Though without any experimental confirmation, the authors concluded the presence of at least two types of active centres.

In regard to the molar ratio between the aluminum alkyl and the transition metal compound influencing MWD, each case should be investigated individually. For example, $(C_5H_5)_2TiCl_2 \cdot AlC_2H_5Cl_2 - Al(C_2H_5)_2Cl$ soluble catalytic system [104] produces, in ethyl chloride, polyethylene having Q = 1.6 at an Al/Ti molar ratio 10 and Q = 3.5 when the ratio is 150.

For polypropylene, it has been found [164] that the MWD of both the crystalline and amorphous fractions, obtained with the α-$TiCl_3 - Al(C_2H_5)_2Cl$ catalytic system, are independent of the Al/Ti ratio. The same result was obtained [143] for polypropylene with $MgCl_2$-supported "high yield" catalytic systems. Instead, with the $(C_5H_5)_2 - TiCl_2 - Al(CH_3)_2Cl$ soluble catalytic system, Chien [164] observed a narrowing of polyethylene MWD with an increase in the Al/Ti ratio and explained it by an increase of chain transfer rate.

In conclusion, electronic density of the transition metal may be influenced, case by case, by the effect of the reaction with aluminum alkyl and, as a result, the carbon-transition metal bond stability, olefin coordination and insertion capacity, stereochemical control of active centre and chain transfer and propagation processes, hence polymer MWD, may also be affected. This is particulary true for soluble catalytic systems for which the existence of active centres as bimetallic complexes is likely.

4.1.5 The Third Component

Many Ziegler-Natta catalytic systems are prepared with the so-called "third component", mostly an electron donor compound [127]. These additives usually show a beneficial effect specially on the stereoregulating capacity of the system. Moreover, they could also exert a certain influence on MWD. As a matter of fact, the third component can modify the catalyst surface through different ways: preferential deactivation of some active centres, activation of others additional species, transformation of some pre-existing centres, complexation and removal of poisons, and complexation with the cocatalyst with possible modification of its composition by means of chemical reactions (a particular type of third component studied [97] is represented by the so called "promoter" widely used with vanadium compounds-aluminum alkyls homogeneous systems). The third component could be therefore a potential tool for MWD regulation, even though the literature does not provide many significant examples on the subject.

Erofeev et al. [128] attributed the polyethylene MWD narrowing observed by adding anisole to $TiCl_4 - Al(C_2H_5)_2Cl$, to a decrease in the number of active centres on the catalyst surface with a consequent decrease in the bimolecular termination and in the low molecular weight fractions. Other russian authors [129] obtained a polyethylene MWD narrowing by adding $Ti(OR)_4$ or alcohol to the same catalytic system used by Erofeev at al. [128].

The use of electron donors is of particular importance for propylene polymerization, where a severe control on polymer stereoregularity is required.

Hirooka et al. [71] found that polypropylene MWD did not vary significantly by adding donors to the $TiCl_3(H)-Al(C_2H_5)_3$ and $TiCl_3(AA)-Al(C_2H_5)_2Cl$ catalytic systems. Instead, Combs et al. [130] found that polypropylene MWD could be narrowed by using catalysts which had been made more stereospecific by adding an electron donor. On the contrary, a patent [132] claims a broader MWD when using electron donor compounds together with the $TiCl_3-Al(C_2H_5)_2Cl$ catalytic system.

While it is often reported [19] that, generally, polypropylene produced with any industrial process presents broad MWD ($Q = 8$–12) with any type of catalytic system, and that in order to obtain a narrow MWD a polymer visbreaking process should be used, studies on the control of MWD by means of "high yield" catalytic systems are being more and more frequently announced. Such systems, developed by Montedison in cooperation with Mitsui Petrochemical [146], are formed of $MgCl_2$-electron donor-$TiCl_4$ and AlR_3.

Suzuki et al. [142] examined the MWD of polypropylene obtained with highly active $TiCl_4-MgCl_2-C_6H_5COOC_2H_5$ and $Al(C_2H_5)_3$. After three hours of polymerization in heptane at 41 °C, the polydispersity index of the integral polymer was 6.5, and those of the isotactic and atactic fractions were 3.8 and 3.7 respectively. These values are lower than those obtained with the conventional $\delta\text{-}TiCl_3-Al(C_2H_5)_3$ catalytic system, though MWD can always be described [143] by the Wesslau's logarithmic normal distribution function [66].

The comparable values of the polydispersity index for both isotactic and atactic polymers led Keii [144] to conclude that the same active centre distribution exists according to the ratio between propagation and transfer rate constants. Furthermore, the low values of the polydispersity index would suggest good active centre homogeneity.

Subsequently, Doi et al. [145] examined other "high yield" supported catalytic systems and, while they confirmed the accomplishment of narrower MWDs than those obtained with non-supported catalytic systems based on $TiCl_4$, they also noticed that polymer polydispersity varied depending on the amount of the Lewis base, the type of aluminum compound used as well as on the type of support.

Kashiwa [148] studied in detail the MWD curve of the boiling heptane insoluble fraction of polypropylene prepared with the $MgCl_2-TiCl_4$ and $Al(C_2H_5)_3-$ $C_6H_5COOC_2H_5$ catalytic system. Such a curve shows two peaks: the position of the lowest molecular weight peak corresponds to the polymer obtained in the absence of ethyl benzoate, while the highest molecular weight peak increases with the increase in the donor/Al ratio causing MWD to broaden. All this, together with the increase in isotactic polymer yield, which is obtained by increasing the donor/Al ratio, led the author to conclude that at least two types of isotactic active centres exist of which the one, producing the polymer with the higher molecular weight, is ester-associated.

With the $MgCl_2-TiCl_3 \cdot 3\ C_5H_5N$ co-milled catalyst treated with $Al(C_2H_5)_2Cl$, Soga [147] found only a slight increase in MWD of the integral polymer, by adding ethyl benzoate to the $Al(C_2H_5)_3$ cocatalyst. However, recent results [198] have shown that, using the "high yield" catalytic systems set up by Montedison, it is possible to significantly vary polypropylene MWD by choosing the appropriate type of donor.

In conclusion, though the addition of a third component to the catalytic system should bring about a MWD narrowing, (both according to diffusion and chemical theory), no effect [113], or noteable MWD broadening have been sometimes observed [242]. These results are probably connected to chemical reactions, with either the active centers or the cocatalyst, which can not be easy foreseen. We belive that, especially since the development of "high yield" supported catalysts, this matter should be more and more investigated.

4.1.6 Support

A considerable improvement in Ziegler-Natta catalysis was achieved starting from the end of the 1960's with the development of the so-called "high yield" catalysts for polyethylene, based on activated $MgCl_2$ [226], which gave rise to the well known advantages in the industrial production of the polymer.

Furthermore the development, during the 1970's, of "high yield" catalysts for polypropylene [146], made it possible to achieve also a better control of polymer stereoregularity. All this was the result of a careful choice, case by case, of an appropriate combination among carrier, transition metal compound, metal alkyl and possible third component.

In particular, the chemical-physical nature of the carier must guarantee not only the dispersion and the optimal utilization of the active species, but also a controlled modification of the kinetic constants of the elementary stages of the polymerization as a consequence of the new electronic environment of the transition metal. Some excellent reviews have been published on this subject [2, 11, 16, 250].

In recent years, especially russian researchers [5], among others, made a great effort in determining the number of active centres and the relevant kinetic constants of such supported catalytic systems.

Ivanchev et al. [133] compared, under the same ethylene polymerization conditions, the performances of some supported systems with those of traditional ones. The authors not only found that the activity of supported systems is determined by the chemical-physical characteristics of the carrier, which influence the number and the propagation constant of the active centres, but also, in agreement with other authors [134], that changes take place, in the elementary polymerization processes, such as to modify the polymer molecular structure. Indeed, as one can observe (Table 6), it is

Table 6. Molecular weights and polydispersity of high density polyethylene obtained with some traditional and supported Ziegler-Natta catalytic systems[a]

Catalytic system	Polymn. Temp. (°C)	Hydrogen	$\bar{M}_n \times 10^{-3}$	$\bar{M}_w \times 10^{-3}$	Q	Ref.
$\alpha-TiCl_3 \cdot 0.3\ AlCl_3$	70	no	340	2700	8.0	[133]
$TiCl_4$	70	no	180	2160	12.0	[133]
$TiCl_4$ on activated carbon	70	no	300	5400	18.0	[133]
$TiCl_4$ on silica-alumina	70	no	300	2990	10.0	[133]
$TiCl_4$ on MgO	70	no	1000	2920	2.9	[133]
$TiCl_4 + Mg(OC_2H_5)_2$	85	yes	17	120	7.0	[139]
$TiCl_4/MgCl_2$ comilled	75	yes	15	93	6.2	[209]

[a] Polymerization in hydrocarbon slurry with $Al(C_2H_5)_3$ as cocatalyst.

possible to regulate polyethylene MWD by changing the nature of the carrier. According to the russian authors, it is the chemical nature of the carrier, rather than its physical structure, which influences the reactivity of the surface active species by two possible mechanisms:

a) influence of the carrier ligands on the active titanium-carbon bond;

b) stabilization of the transition metal oxidation state.

The second hypothesis is mainly considered important since it has been observed [133, 135, 236] that, for catalytic systems of the $MgO-TiCl_4$ and $Al(C_2H_5)_3$ type, the number of titanium oxidation states is lower than that found with either aluminum silicate -$TiCl_4$ and $Al(C_2H_5)_3$ or with non-supported catalytic systems. This is due to the retarding effect of the magnesium compound on the transition metal reduction by the aluminium alkyl.

The importance of the carrier was also emphasized by Böhm [11] who found that a supported system, of the $Mg(OC_2H_5)_2-TiCl_4$ and $Al(C_2H_5)_3$ type, yields polyethylene having a narrower MWD ($Q = 7$) than that obtained with the traditional $TiCl_3-Al(C_2H_5)_3$ catalytic system ($Q = 10$). However, in both cases, MWD can be described [139] by the normal logarithmic distribution function [66]. Böhm [237] subsequently also showed that 1-butene/ethylene copolymers, obtained with the same supported catalytic system, present a polydispersity which can be compared to that of polyethylene.

Polyethylene with an even lower degree of polydispersity was recently obtained by Greco et al. [141] by studying "high yield" magnesium chlorotitanate-based catalytic systems [248]. Such results have been explained by a limited number of active species due to a certain homogeneity caused by the magnesium ions. An alternative explanation states that the originally heterogeneous catalyst should work, during the polymerization process, in a *pseudo-homogeneous* soluble form.

Yermakov et al. [5], on the basis of both their own experimental data and the few one available from the literature, concluded that polyethylene, obtained with magnesium-based catalytic systems, though showing properties similar to that produced with traditional Ziegler-Natta catalytic systems ($TiCl_4$-aluminum alkyls), presents generally a narrower MWD.

Finally, a particular case is represented by the $[(C_6H_5)_3SiO]_2CrO_2$-silica supported and AlR_2OR catalytic system [138], which gives polyethylene having a bimodal MWD curve with considerable low and very high molecular weight fractions.

In conclusion, though the carrier is a further possibility in order to control polymer MWD, only a better knowledge of the relationship which link the catalyst surface to the number, composition and structure of active centers (hence to the rate constants of the elementary stages of the polymerization process) should permit an even greater control on polymer molecular properties and, in particular, on polymer MWD.

4.2 Polymerization Parameters

The study of MWD as a function of the main polymerization parameters can be a useful tool of investigation to verify the agreement between proposed theories and mechanisms and experimental data. Practical interest in such study, in turn, derives from the possibility, at least potential, of obtaining variations in MWD and, therefore, in polymer characteristics by acting on the process parameters.

4.2.1 Temperature

MWD is usually broadened with a decrease of polymerization temperature. In the ethylene polymerization with the $(C_5H_5)_2TiCl_2-Al(CH_3)_2Cl$ soluble catalytic system, Chien [164] attributed such behaviour to a greater decrease in termination rate in comparison to propagation rate as temperatures decreases.

Similar dependence of MWD on temperature was also observed for polypropylene both by Chien [164] using the α-$TiCl_3-Al(C_2H_5)_2Cl$ catalytic system and by Combs et al. [130] with the α-$TiCl_3-AlC_2H_5Cl_2$-hexamethylphosphoric triamide system. Such an influence was attributed to a number of causes: greater frequency of chain transfer processes at higher temperature; levelling off in specific activity of the different types of active centres or progressively less blocking of monomer transport with an increase in temperature. More recently the same behaviour has been found by Yuan et al. [242] with the $TiCl_3(AA)-Al(C_2H_5)_2Cl$ catalytic system.

Rishina and Vizen [176] observed that the MWD of the atactic fraction of polypropylene prepared with the δ-$TiCl_3-Al(C_2H_5)_3$ catalytic system narrows when polymerization temperature increased from 20 to 70 °C. This phenomenon was ascribed to a greater reactivity of the active centres for growing chain termination.

A slight narrowing of the integral polymer MWD was also observed [225], with some "high yield" catalysts for polypropylene, with increasing temperature (Q went from about 7.5 to about 6.0, going from 60 to 80 °C).

A peculiar behaviour was recently found [166] in the case of ethylene polymerization in heptane with $Ti(OR)_nCl_{4-n}-MgCl_2$ and $Al(C_8H_{17})_3$ "high yield" catalytic system in the temperature range from 50 to 200 °C. The polymer MWD, which was broad at low temperature, narrowed to a minimum, at about 100–120 °C, to broaden again at higher polymerization temperatures. MWD broadening, above the polymer solubility temperature, is difficult to explain on the basis of diffusion theories and can be more likely ascribed to the formation of new active species possibly by transformation of the pre-existing ones.

4.2.2 Hydrogen

According to the theory of Gordon and Roe [86], the introduction of a chain transfer agent during polymerization should cause MWD to narrow. Instead, Berger et al. [7] proved that, for polyethylene obtained with traditional Ziegler-Natta catalytic systems, the Q value is independent of the hydrogen concentration, thus being in favour of the active sites variety model as for the MWD origin.

Taylor and Tung [167] reached the same conclusions for polyethylene prepared with $TiCl_4-Al(i$-$C_4H_9)_3$, Keii [144] for polypropylene prepared with $MgCl_2$-supported "high yield" catalytic systems, and Böhm [139] for polyethylene obtained with $Mg(OR)_2$-based "high yield" catalytic systems.

Recently Yuan et al. [242] reported that polypropylene, prepared with $TiCl_3 \cdot 1/3AlCl_3$ (Stauffer type AA)$-Al(C_2H_5)_2Cl$, showed slightly higher polydispersity with hydrogen than without hydrogen.

On the other hand, by fractionating isotactic polypropylene prepared in the presence of hydrogen with α-$TiCl_3-Al(C_2H_5)_3$, Pegoraro [68] found polydispersity values lower than those observed by other researchers [67] in the absence of hydrogen.

Even with some "high yield" catalytic systems, the index of polydispersity was inversely proportional to the hydrogen concentration[137].

The living polymerization of propylene at -78 °C with the $V(acac)_3-Al(C_2H_5)_2Cl$ soluble catalytic system [112] is a special case. In this instance hydrogen, acting as a chain transfer agent only, brings the index of polydispersity ($Q = 1.05-1.20$) from the theoretical value in the absence of transfer, to values near the theoretical one ($Q = 2$) according to Schulz-Flory. This is due to an increase in the number of polymeric chains per atom of vanadium and to a corresponding decrease in $\bar{M}_n$.

Results concerning MWD evaluation in relation to molecular weight or polymer viscosity are also contradictory. According to some authors [66,72,167] polydispersity depends on average molecular weight. However, Yamaguchi [77], by studying polypropylene MWD with the column elution method, has demostrated that this is not true and that some results [71], were not correct since MWD data were treated according to the conventional log-normal function by neglecting peculiar patterns at the high molecular weight side of the distribution curves observed.

4.2.3 Time

In principle, if polymerization under *pseudo-stationary* conditions is kinetically controlled and the active centres decay spontaneously, the polydispersity is independent of either yield or polymerization time [55]. Therefore, if polymerization time is found to influence MWD, this may depend on the lack of one or more of the mentioned conditions or on eventual diffusion phenomena.

An example of how the variation of MWD with time can constitute a useful mean to determine or confirm the polymerization mechanism, was provided [164] in the case of ethylene polymerization with the $(C_5H_5)_2TiCl_2-Al(CH_3)_2Cl$ soluble catalytic system at 15 °C and Al/Ti ratio of 2.5. Polymer polydispersity was decreasing during the very first moments of polymerization reaching a minimum, corresponding to a maximum growing chain concentration, and then increased with time during the pseudo-stationary state. Such an increase is in agreement with the proposed kinetic scheme which requires a first order initiation and a termination due to biomolecular disproportionation.

Belov et al. [54] experimentally demonstrated that with the $(C_5H_5)_2TiCl_2-Al(C_2H_5)_2-Cl$ homogeneous catalytic system, the degree of polyethylene polydispersity reaches the theoretical value 2 in a short polymerization time. This agrees with a kinetic model based on monomolecular reactions of chain transfer and termination.

Instead, with another homogeneous catalytic system, namely $(C_5H_5)_2TiC_2H_5Cl-AlC_2H_5Cl_2$, the polyethylene polydispersity increases when time is increased [163]. Such a result was attributed to the type of non-stationary kinetic in which the rate decay is connected to an irreversible active centre deactivation rather than to a diffusion type hindrance. Similar behaviour for ethylene polymerization with $(C_5H_5)_2TiCl_2-Al(C_2H_5)_2Cl$ catalytic system, was previously ascribed, by Belov et al. [103], to the growth with time of a new kind of active centres as experimentally proved.

With traditional heterogeneous Ziegler-Natta catalytic systems, polymers with broad MWD are formed just during the first instants of polymerization [65]. At 25 °C with $TiCl_4-Al(i-C_4H_9)_3$, polyethylene MWD tends to broaden in the first thirty minutes of polymerization, in both the very low and the very high molecular weights fractions, probably owing to a non-stationary state, while the maximum position of the distribution curve remains unchanged [167].

Bier et al. [174] found that in propylene polymerization, at 50 °C with aluminum alkyl reduced $TiCl_3-Al(C_2H_5)_2Cl$, the polymer polydispersity increased (Q from 6 to 10 in the first 30 minutes) and then decreased with time (Q approximately 4). Such a phenomenon could be justified by a decrease in the number of active centres with time after an initial period of rapid increase, together with the living nature of the polymeric chain which is only terminated by monomolecular disproportionation [172,173].

Using $TiCl_3(AA)-Al(C_2H_5)_2Cl$ in the polymerization of both ethylene and propylene, Schnecko et al. [170] found initially high polydispersity values which then slowly decrease with time. In this case, the variation of the chain transfer rate with the cocatalyst was held responsible for the phenomenon. This variation tends to have less influence as conversion increases with the consequent increase in $\bar{M}_n$. Instead, for butene-1 polymerization, it was observed that the degree of polydispersity, which was initially low, increased with time corresponding to an increase in the chain transfer and a decrease in $\bar{M}_n$. The differences in behavior among the monomers were attributed to differences in propagation constants for active centers differently distributed on the catalytic surface and also to physical phenomena which may differently affect the kinetic behavior of both propagation and chain transfer processes (different polymer solubility and morphology in the reaction medium depending on which monomer is considered) [171].

From an interpretative point of view the behavior of highly active catalytic systems supported on magnesium compounds is equally complicated.

Crabtree et al. [169] studied ethylene polymerization in the presence of hydrogen with the $RMgX-TiX_4$ and AlR_3 catalytic system (R = alkyl, X = halogen). In agreement with the proposed polymerization mechanism based not only on spontaneous active centres decay, but also on a diffusive control, especially in the first moments of the polymerization it was found that initial extremely high molecular weights are formed. The MWD is broad and then narrows with polymerization to an asymtoptic value. A somewhat similar result, justified by a different mechanism, was found by Böhm [139,140] in a detailed kinetic study of ethylene polymerization with $Mg(OC_2H_5)_2-TiCl_4$ and AlR_3 at 85 °C. The author found that, under appropriate experimental polymerization conditions, there are no diffusion obstacles to monomer transfer; thus a very high concentration of active centres together with the porous nature of polymer particles give rise to very high yields. In the absence of hydrogen, the transfer processes with cocatalyst and with monomer (β-elimination) have approximately the same importance. This characteristic and the kinetic data obtained in relation to the value of the propagation constant make one consider the behaviour of the catalytic system as similar to that of traditional systems with one difference connected to the greater number of its active centres. Based on the experimental results relevant to the changing of $\bar{M}_n$ and Q values with time (the first initially grows and then becomes constant, while the second decreases from an initially high value to a constant one, Q = 7.5 ± 2), the author deduced that there are some active centres characterized by a much higher propagation constant than the global one, but that they are able to produce, due to a very short average life time, macromolecules with extremely high molecular weights only in the first moments of polymerization.

MWD variation of polypropylene with time has been investigated by many authors. Doi et al. [108] found no dependence of MWD on polymerization time (in the range 2 to 10 hours) for syndiotactic polypropylene obtained at $-78\ °C$ with $VCl_4-Al(C_2H_5)_2Cl$ homogeneous catalytic system.

For the crystalline polymer obtained with $\alpha\text{-}TiCl_3-Al(C_2H_5)_2Cl$ at Ti/Al ratio 1, a slight decrease in the index of polydispersity was found [164] between 0 and 3000 minutes. This has been ascribed both to possible diffusion phenomena, even for low conversions, and to the kinetic behaviour of two different types of active centres.

Vizen and Yakobson [189] studied by G.P.C. the isotactic fraction of the polypropylene obtained at 30 °C with the $\delta\text{-}TiCl_3-Al(C_2H_5)_2Cl$ catalytic system. With this catalytic system, after the first ten minutes of growth, the polymerization rate becomes stationary for approximately an hour and, in the absence of chain transfer, $\bar{M}_n$ increases linearly with conversion. These conditions, of slow initiation and a subsequent period of stationary state, should lead to a theoretical calculated value Q decreasing from 1.25 to 1.11 from the eighth to the fortieth minute of polymerization, provided that the active centres were of the same type. In reality, Q was found to decrease from 5 to 2.5. Since polymerization conditions were such as to exclude diffusion phenomena, the experimental MWD data were justified from a mathematical model based on the only heterogeneity of the active centres with respect to propagation rate constants. According to the authors, four is the minimum number of active species necessary to justify both the numeric value and shape of the distribution curve. The same model preducts instead that, for polymerization with hydrogen, MWD will be independent of time, as experimentally verified.

With a lot of work [99,112,131,177] on the "high yield" $MgCl_2-TiCl_4-C_6H_5COOC_2H_5-Al(C_2H_5)_3$ catalytic system, et al. found [142] that polypropylene MWD, both of the integral polymer and of the boiling heptane soluble and insoluble fractions, was scarcely dependent on polymerization time. With this catalyst system $\bar{M}_n$ did not vary with time and the chain transfer process has been operating right from the first moments of polymerization; the transfer constant would result much greater than that of the traditional catalytic systems, thus balancing the higher propagation constant.

Also other authors [238] have verified that the MWD of polypropylene obtained with $MgCl_2$-supported "high yield" catalytic systems is independent of polymerization time.

4.2.4 Other Parameters

The importance of other parameters, such as monomer and catalyst concentration, conversion etc., on MWD must be examined in relation to the type of catalytic system being considered and, above all, to incidental phenomena connected to mass or heat transfer.

Meyer and Reichert [168] found that, for polyethylene prepared at 80 °C with the highly active $Mg(OC_2H_5)_2-TiCl_4$ and AlR_3 catalytic system, the MWD is practically independent of conversion and of both catalyst and polymer particle size, yet it broadens with an increase in catalyst concentration.

Similar catalyst concentration dependence has been found for soluble systems and it has been explained by a kinetic model [164].

Studying the influence of operating conditions on the MWD of polypropylene obtained with three-component catalytic systems, Combs et al. [130] observed that it narrows at high monomer concentrations or at low conversions, especially for low monomer concentrations.

For ethylene polymerization in the presence of hydrogen with the highly active magnesium alkyl-reduced $TiCl_4$ and $Al(C_8H_{17})_3$ catalytic system, Crabtree et al. [169] observed that the index of polydispersity decreased with an increase in conversion reaching the asymptotic value 4, corresponding to a yield of 10 kg of polymer per gram of $TiCl_3$ at atmospheric pressure. This agrees with a model which, along with a gradual disappearance of active centres, predicts also limitations due to diffusion control, especially in the first moments of polymerization.

Recently Yuan et al. [242] found a dependence of MWD on the solubility parameter of the medium in the slurry polymerization of propylene over $TiCl_3(AA)-Al(C_2H_5)_2Cl$.

This review covering the influence of polymerization parameters on MWD clearly shows how impossible it is to elaborate laws valid for all catalytic systems. Furthermore, numerous discrepancies among the experimental data prevent any attempt at their rationalization based on a general kinetic model, even for similar catalytic systems. However, at least qualitatively, the following conclusions can be reasonsably drawn:

— the increase in hydrogen concentration tends to narrow MWD or, at least, to leave it unaltered;
— the increase in polymerization temperature tends to narrow MWD (Table 7);
— the increase in polymerization time and conversion tends to bring the index of polydispersity to an asymptotic value, often the lowest possible for a given catalytic system (Table 8).

Table 7. Influence of temperature on polydispersity expressed as Q index of the polymer depending on the catalytic system

Temperature (°C)	$(C_5H_5)_2TiCl_2-Al(CH_3)_2Cl$[a, d]	$\alpha-TiCl_3-Al(C_2H_5)_2Cl$[a, e]	$\alpha-TiCl_3-HPT-AlC_2H_5Cl_2$[b, e]	$TiCl_3(AA)-Al(C_2H_5)_2Cl$[c, e]
0	6.1	—	—	—
15	5.4	—	—	—
30	4.5	—	—	6.41
40	—	—	8	—
45	3.6	—	—	—
50	—	4.5	—	5.92
60	2.4	—	3	—
70	—	—	2	5.79
80	—	—	2	—
90	—	3.4	—	5.62

[a] From Ref. [164]; [b] Estimated from Ref. [130]; [c] From Ref. [242]; [d] Polyethylene; [e] Polypropylene

Table 8. Influence of polymerization time on polydispersity expressed as Q index of the polymer depending on the Catalytic system

Time (minutes)	$TiCl_3$-Al(C_2H_5)$_2$Cl[a] polypropylene	$TiCl_3$-Al(C_2H_5)$_3$[a] polypropylene	$TiCl_3$(AA)-Al(C_2H_5)$_2$Cl[b]		α-[c] and δ-$TiCl_3$-Al(C_2H_5)$_2$Cl[d] polypropylene	$TiCl_4$/MgCl$_2$-C_6H_5COOC$_2$H$_5$-Al(C_2H_5)$_3$[e] polypropylene
			polypropyl.	polyethyl.		
0.25						5.3
0.75						5.2
4						5.2
8						
10	6.6	5.4	11	24	6	
30	10.7	8.5	11	21	4	
40					3	5.5
50					2.5	
60	6.7	7.8	10	20		
90	6.3	6.7	10	18		
100					8	
120	6.9	4.8	9	18		
150	4.2	5.3				
180					2.5	6.5
500					7	
1000					6	
2000					5	
3000					5	

a) From Ref. [174]; b) Extimated from Ref. [170]; c) Extimated from Ref. [164]; d) Extimated From Ref. [189]; e) From Ref. [142]

5 Considerations about MWD Control of Polyolefins in Industry

5.1 Polymerization Processes

In industrial production many factors determine the choice between the many possible polymerization processes [14, 15, 24, 25]. Among these factors one must mention efficiency, easy control of the reaction, flexibility, process economy, closely connected with the chosen catalytic system on the one hand and with the desired key properties of the polymer on the other hand, as well as with the characteristics of size, shape and density of the polymer particles (Table 9).

For general theories on the dependence of polymer MWD on the type of reactor, whether continuous or discontinuous, and on the operating conditions, based on assumed kinetic mechanisms, one must refer to specialized literature [211-214].

Concerning the polymerization with Ziegler-Natta systems, McGreavy [205] attempted to correlate some design parameters of a continuous stirred tank reactor (CSTR) for a slurry process, at least with the most important *moments* of the polymer MWD. However, the complexity of the physical and chemical phenomena and the uncertainty as to the experimental values of some parameters did not allow him to identify any precise correlation. A method for predicting the MWD of polyethylene synthesized in a continuous reactor with Phillips-type catalyst, from batch reactor data, has been given by Hoff and Shida [210].

In the continuous stirred tank reactor, a considerable MWD enlargement should be expected [243], by intentionally cycling the concentration of chain transfer, expecialiy for fast cycling and imperfect mixing.

Thus an appropriate choice of the operating conditions, for the most part not reported in the literature as this is the subject of specific know-how, allows one to optimize the polymer properties which are, however, fundamentally predetermined by the type of catalytic system employed.

Polyethylene (for which MWD control is particularly necessary) is usually produced in one of the following techniques:
a) slurry process;
b) solution process;
c) vapor phase process.

The three processes differ by the operating conditions as reaction temperature, type of heat control and degree of system heterophasicity. Moreover, different processes could bring about also different thermal effects inside the polymer particle [254]. The solution process mainly yields polyethylene with a medium to very narrow (Q values of 2 or 3) MWD, while both slurry and vapor phase [114-115] processes give the whole range of products with narrow to broad MWD.

5.2 Patent Literature Survey

The great industrial importance of the polymerization of olefins with Ziegler-Natta catalysis has stimulated a lot of research, on a more or less empirical basis in the major manufacutring companies and still does today.

For this reason any approach toward the understanding of Ziegler-Natta catalysis cannot neglect a careful examination of the patent literature. The well known complexi-

Table 9. Main parameters of industrial Ziegler-Natta polymerization which could affect polyolefins MWD

Catalyst	Cocatalyst	Polymerization Process	Polymerization Parameters	Industrial Requirements
– Physical state (homo-geneous, heterogeneous) and physcial structure (crystallite size, porosity etc.) – Type of transition metal atom – Type of carrier (if any) – Electronic and steric environment of the metal atom (type of ligand, electron donor etc.)	– Type of metal – Chemical and structural nature of ligands – Concentration in the reaction medium and stoichiometric ratio to the transition metal atom – Type of eventually present electron donor	– Type (slurry, solution, vapor phase) – Mono- or multi-stage	– Temperature – Residence time – Catalyst, monomer and polymer concentration – Molecular weight regulator – Mixing degree – Comonomer (if any) – Reaction medium	– Catalytic activity – Linearity (HDPE) – Stereoregularity (PP) – MW regulation – Polymer powder morphology and particle size distribution – Process economicity, simplicity and flexibility

Table 10a. High density polyethylene MWD: Catalyst control based on the transition metal compound. Selected collection from patent literature since 1968. Results are given for one of the typical patent examples. Unless otherwise stated, polymerization is carried out in batch. Polymerization temperature generally ranges from 70 to 90 °C for the slurry and gas phase processes, whereas it is above the polymer melting point for the solution process. Molecular weight is controlled almost exclusively by hydrogen

Patent Application (Priority)	Company	Catalyst	Cocatalyst	Process[a]	Polymer MWD[b]	Remarks[b]
Ger. Offen. 2, 140, 326 (1970)	Mitsui Petro-chem.	$MgO + Et_3Al/VCl_4/TiCl_4$	Et_3Al	SO	Q12 to 20	Q 10 without VCl_4; MWD control by changing V/Ti ratio
U.S. 3, 752, 795 (1972)	Standard Oil	$AlCl_3 + Mg(OH)_2/CrO_3 + Ti(OBu)_4$	Et_3Al	SL	Q 7	MWD may be varied by changing Cr/Ti ratio
Ger. Offen. 2, 528, 558 (1974)	Mitsubishi Chem. Ind.	$TiCl_4 + VOCl_3 + VO(OR)_3/Et_3Al_2Cl_3$	Et_3Al	SL	FR 24 to 37	MWD control by varying Ti/V ratio
Ger. Offen. 2, 543, 437 (1974)	Solvay	$BuOH + J_2 + Mg + Ti(OBu)_4/ZrCl_4/EtAlCl_2$	$i\text{-}Bu_3Al$	SL	FR up to 121	FR 59.5 without $ZrCl_4$; FR 66.9 without $Ti(OBu)_4$ Polymerization temp. 60 °C.
Japan Kokai 78 44, 495 (1976)	Asahi Chem. Ind.	$Mg(OEt)_2 + SOCl_2/TiCl_4/AlMg_6Et_3Bu_{12}/TiCl_4$	$i\text{-}Bu_3Al$	SL	FR 76	FR 45 without the last treatment with $TiCl_4$
Belg. 871, 221 (1977)	Naphtachimie	$Mg + BuCl + TiCl_4 + Ti(Oi\text{-}Pr)_4$	$Octyl_3Al$	GP	Q 3	Q 5 without $Ti(Oi\text{-}Pr)_4$
Ger. Offen. 2, 925, 949 (1978)	Chisso	$Mg(OH)_2 + AlCl_3/TiCl_4 + $ dimethylpolysiloxane$/TiCl_4 + Ti(Oi\text{-}Bu)_4$	$i\text{-}Bu_3Al$	SL	Q 32	Q 5 without $TiCl_4 + Ti(Oi\text{-}Bu)_4$
Eur. Pat. Appl. 45, 885 (1980)	Montedison	$MgCl_2 + Ti(OBu)_4/TiCl_4$	$i\text{-}Bu_3Al$	SL	MI_{10}/MI 9.6 to 13.3	MI_{10}/MI 8.5 by using $SiCl_4$ instead of $TiCl_4$

a) SL = Slurry process; SO = Solution process; GP = Gas phase process;

b) MI = Melt index (ASTM D 1238, condition E); HLMI = Flow rate (ASTM D 1238, condition F); MI_{10} = Melt Index (ASTM D 1238, condition N);
FR = Flow ratio: ratio of melt index HLMI to melt index MI

Table 10b. High density polyethylene MWD: Catalyst control based on the carrier

Patent Application (Priority)	Company	Catalyst	Cocatalyst	Process[a]	Polymer MWD[b]	Remarks[b]
U.S. 3, 644, 318 (1968)	Hoechst	$Mg(OEt)_2$ + $TiCl_2(O\ i\text{-}Pr)_2$	Et_3Al	SL	Q 2.9	By adding $Mg(OH)_2$ to the catalyst, Q value can be controlled from 4 to 8 by the content of hydroxy groups
U.S. 3, 879, 368 (1971)	Union Carbide	SiO_2 + $CrCp_2$/Et_3Al/ $MeEt_2SiH$		GP	FR 35 to 65	The lower the activation temperature of the support, the broader is the MWD
Fr. Demande 2, 207, 931 (1972)	Solvay	γ-Al_2O_3 + $MgCl_2$ · $4H_2O$/$TiCl_4$	$i\text{-}Bu_3Al$	SL	$(\bar{M}_z/\bar{M}_w)$ – 111 to 28	Polydispersity is function of the Mg content
Fr. Demande 2, 271, 236 (1974)	Nippon Oil	mixture of (A) (Al_2O_3 + SO_3/$TiCl_4$) and (B) ($MgCl_2$ + $Al(OEt)_3$ + $TiCl_4$)	Et_3Al	SL	$\log \dfrac{HLMI}{MI}$ 1.97	For catalysts containing only (A) or (B), flow parameter is 1.87 or 1.55 respectively
Ger. Offen. 2, 621, 591 (1975)	Nissan Chem. Ind.	BuMgCl + polymethyl-hydrosiloxane/hydrotalcite/$TiCl_4$	$i\text{-}Bu_3Al$	SL	MI_{10}/MI 22	Flow ratio 8.8 without hydrotalcite
U.S. 4, 128, 502 (1976)	Standard Oil	$AlCl_3$ + $Ti(OBu)_4$/ $Mg(OH)_2$/$EtAlCl_2$	Et_3Al	SL	FR 57.7	FR 33.2 with $Mg(OEt)_2$ instead of $Mg(OH)_2$
U.S. 4, 198, 315 (1978)	Dow Chem.	$NiCl_2$ + Bu_2Mg l/6 Et_3Al/$EtAlCl_2$/Et_3Al/ $Ti(Oi\text{-}Pr)_4$		SO	Narrow to broad	MWD control according to catalyst composition; narrower MWD without $NiCl_2$

[a, b] Conditions and foot note, see Table 10a

Table 10c. High density polyethylene MWD: Catalyst control based on a third added component

Patent Application (Priority)	Company	Catalyst	Cocatalyst	Process[a]	Polymer MWD[b]	Remarks[b]
Belg. 744, 522 (1969)	Solvay	$MgO + Et_3Al/TiCl_4$	$i\text{-}Bu_3Al$	SL	Broad	Narrower MWD without Et_3Al
U.S. 3, 859, 267 (1972)	Mitsubishi Chem. Ind.	$VOCl_2 + TiCl_n(OR)_{4-n} + H_2O/TiCl_4/R_mAlCl_{3-m}$	$i\text{-}Bu_3Al$	SL	FR 27 to 51	MWD control by varying the water amount
Ger. Offen. 2, 635, 550 (1975)	Chisso	$Mg(OH)_2 + AlCl_3/TiCl_4/ Al(Oi\text{-}Pr)_3/TiCl_4$	Et_3Al	SL	Q 20.4	Q 7.5 without $Al(Oi\text{-}Pr)_3$
U.S. 4, 238, 355 (1978)	Dow Chem.	$R_2Mg + HCl + Ti(Oi\text{-}Pr)_4 + Et_2Zn$		SO	MI_{10}/MI 15.80	Flow ratio 10.97 without Et_2Zn

a, b) Conditions and footnotes see Table 10a

Table 10d. High density polyethylene MWD: Catalyst control more complex or not given

Patent Application (Priority)	Company	Catalyst	Cocatalyst	Process[a]	Polymer MWD[b]	Remarks[b]
Brit. 1, 258, 984 (1968)	Hoechst	$Mg(OAc)_2$ + $TiCl_2(Oi\text{-}Pr)_2$	Et_3Al	SL	Q 3.4	
Japan. Kokai 73 66, 179 (1971)	Mitsui Petrochem.	$Mg(OH)_2Al(OH)_3 \cdot 4\,H_2O$ + $MgCl_2/TiCl_4$	Et_3Al	SO	Q 26.6	
Ger. Offen. 2, 135, 044 (1971)	Veba Chemie	SiO_2 + CrO_2Cl_2	Et_3Al	SL	$(\bar{M}_w/\bar{M}_n) - 1$ 30.2	
Fr. Demande 2, 243, 208 (1973)	Solvay	Mg + ROH + $Ti(OBu)_4$/$EtAlCl_2$	$i\text{-}Bu_3Al$	SL	Narrow	
Fr. Demande 2, 266, 706 (1974)	Mitsubishi Chem. Ind.	$Mg_{0.8} \cdot Ti_{0.2}Cl_{2.2}THF_{1.2}$	$i\text{-}Bu_3Al$	SL	Q 2.9	
Ger. Offen. 2, 615, 390 (1975)	Solvay	$Mg(OR)_2$ + $Ti(OR)_4$ + $Zr(OR)_4$/$RAlCl_2$	$i\text{-}Bu_3Al$	SL	FR 37 to 142	
Ger. Offen. 2, 822, 809 (1977)	Montedison	$MgCl_2$ + $Ti(OBu)_4$/$SiCl_4$ + polymethyl-hydroxiloxane	$i\text{-}Bu_3Al$	SL	MI_{10}/MI 8.0	
Japan. Kokai 78 88, 891 (1977)	Mitsui Toatsu Chem.	Hydrotalcite + $SnCl_4$/$TiCl_4$	$i\text{-}Bu_3Al$	SL	MI_{10}/MI 16	
Eur. Pat. Appl. 19, 637 (1978)	Mitsubishi Chem. Ind.	Organic oxidized or halo-genated compounds of Ti, V, Zr or Hf/Al RX_2	R_3Al	SL	Broad	
Ger. Offen. 2, 802, 819 (1978)	Hoechst	ter-BuOH + CrO_3 + SiO_2/$Ti(Oi\text{-}Pr)_4$	poly-i-Bu Alu-moxane	SL	Q 20.5	
Ger. Offen. 3, 010, 202 (1980)	Wacker Chemie	sec-BuMgBu + $CHCl_3$/$Al(OR)_3$ + $TiCl_4$/$TiCl_4$ + $Et_3Al_2Cl_3$	$Octyl_3Al$ or $Isoprenyl_3Al$	SL	Broad	
Japan. Kokai Tokkyo Koho 81 53, 109 (1979)	Sumi tomo Chem.	(A) BuMgCl + MeOH/(B) toluene + Et_2AlCl + Bu_2O + $TiCl_4$/(C) $TiCl_4$	$EtAl(OEt)_2$ + Et_2AlCl	SL	FR 92	FR 31 for a catalyst with (A) and (C) only

a, b) Conditions and foot notes see Table 10a

Table 11. High density polyethylene MWD: Control by the cocatalyst

Patent Application (Priority)	Company	Catalyst	Cocatalyst	Process[a]	Polymer MWD[b]	Remarks[b]
Brit. 1, 332, 510 (1970)	Union Carbide	$SiO_2 + (R_3Si)_2CrO_4$	mixture of two $R_xAl(OR')_{3-x}$	SL or GP	Narrow to broad	MWD control depending on the type of hydrocarbyl aluminium alkoxides mixture
Ger. Offen. 2, 128, 760 (1970)	Shell Int. Res.	$BuMgCl \cdot Bu_2O + TiCl_4$	$(C_nH_{2n+1})_3-Al$	SL	$\log \frac{HLMI}{MI}$ 1.8 to 2.4	MWD broadens by increasing chain length of trialkylaluminium ($n = 8 - 18$)
Ger. Offen. 2, 160, 431 (1970)	Shell Int. Res.	$BuMgCl \cdot Bu_2O + TiCl_4$	$Octyl_3Al. + Et_2AlCl$	SL	$\log \frac{HLMI}{MI}$ 1.94	Flow parameter 1.77 without Et_2AlCl
Ger. Offen. 2, 519, 071 (1974)	Asahi Chem. Ind	$AlMg_6Et_3Bu_{12} + TiCl_4$	$i\text{-}Bu_3Al + Cl_2CHCH_2Cl$	SL	Q 20	Q 7.8 without Cl_2CHCH_2Cl
Ger. Offen. 2, 523, 098 (1974)	Asahi Chem. Ind	$TiCl_4 + Et_{2.4}Al(OEt)_{0.6}$	$EtAlCl_2 + Et_{2.02}Al(OEt)_{0.98}$	SL	Q 8.7	Q 14.2 without $Et_{2.02}Al(OEt)_{0.98}$
Ger. Offen. 2, 755, 192 (1977)	Chem. Werke Hüls	$VOCl_3 + PrOH/EtAlCl_2$	$i\text{-}Bu_3Al$	SL	$(\bar{M}_w/\bar{M}_n) - 1$ 22.2	Inhomogeneity 5.5 by using $EtAlCl_2$ instead of $i\text{-}Bu_3Al$
Eur. Pat. Appl. 9, 426 (1978)	Naphtachimie	$Mg + J_2 + TiCl_4 + BuCl$	$Octyl_3Al + vinyl\text{-}chloride$	SL	Q 8.7	Q 6.1 without vinylchloride
Jpn. Kokai Tokkyo Koho 79 134, 792 (1978)	Mitsui Toatsu Chem.	$MgCl_2 + EtOH/Et_2AlCl/ TiCl_4$	$i\text{-}Bu_3Al + AlCl_3$	SL	Q 15.1	Q 8.4 without $AlCl_3$
Jpn. Kokai Tokkyo Koho 81 99, 206 (1980)	Idemitsu Kosan	$Mg(OEt)_2 + SiCl_4/ i\text{-}PrOH/TiCl_4$	$Et_3Al + I_2$	SL	FR 65	FR 38 without I_2

[a, b] Condition and footnotes see Table 10a

Table 12. High density polyethylene MWD: Control by the polymerization process

Patent Application (Priority)	Company	Catalyst	Cocatalyst	Process[a]	Polymer MWD[b]	Remarks[b]
Belg. 828, 598 (1974)	Monsanto	$VOCl_3$ supported on polyethylene	Et_3Al + Et_3Al_2-$(OEt)_3$	GP	Q 5.9–13.3	Q 13.3 at polymn. temperature of 85–90 °C. Q 5.9 at polymn. temperature of 110–115 °C
U.S. 3, 957, 448 (1974)	Standard Oil	Titanium compound	Et_3Al	GP	Q 9	Polymn. effected in a 2-zone reactor; Q 6 for a polymn. process in a reactor with a single compartment
Ger. Offen. 2, 635, 273 (1975)	Stamicarbon	$EtAlCl_2$ + $Bu_2Mg/TiCl_4$		SL	Very broad	Polymn. process with a 2-steps loop reactor
Ger. Offen. 2, 826, 548 (1977)	Nippon Oil	$MgCl_2$ + $TiCl_4$	Et_3Al	SL	Broad	Continuous 2-stage polymn. in liquid medium at different pressures
Eur. Pat. Appl. 22, 376 (1979)	Mitsui Petro-chem. Ind.	$MgCl_2 \cdot 5$ $EtOH/Et_2AlCl/$ $TiCl_4$	Et_3Al	SL	Very broad	Polym. process conducted in a multiplicity of steps
Eur. Pat. Appl. 24, 881 (1979)	Chisso	$Mg(OH)_2$ + $AlCl_3/TiCl_4$ + dimethylpolysiloxane	Et_3Al	SL	Q 25	Continuous 2-stages polymn.; Q 6 with a single reactor
Belg. 883, 687 (1979)	Sumitomo Chem.	$BuMgCl$ + Bu_2O + $AlCl_3$ or $SiCl_4/TiCl_4$	Et_3Al	SL	Very broad	Polymn. in a multistage process
Belg. 884, 866 (1979)	Asahi Chem. Ind.	$AlMg_6Et_3Bu_{12}$ + $TiCl_4$	Et_3Al	SL	Very broad	FR 180 with a 3-steps polymn. process; FR 45 with a single reactor
Jpn. Kokai Tokkyo Koho 81 38, 303 (1979)	Asahi Chem. Ind.	Bu_2Mg + $Et_3Al/BuOH/$ $SiHCl_3/TiCl_4$	Et_3Al	SO-SL	Very broad	Continuous polymn. process with a 1st soln. polymn. stage and a 2nd slurry polymn. stage
Eur. Pat. Appl. 40, 992 (1980)	Mitsui Petro-chem. Ind.	$MgCl_2$ + $EtOH/Et_2AlCl/$ $TiCl_4$	Et_3Al	GP	Q 28.7	Polymn. in at least 2 indipend. zones; Q 7.7 with a single zone

[a, b] Conditions and footnotes see Table 10a

Table 13. High density polyethylene MWD: Control with mixed systems

Patent Application (Priority)	Company	Catalyst	Cocatalyst	Process[a]	Polymer MWD[b]	Remarks[b]
Ger. Offen. 2, 043, 515 (1970)	Hoechst	$TiCl_3$ with $Mg(OH)_2$/ $Ti(Oi\text{-}Pr)_4$/$TiCl_4$ or $TiCl_3$ with $Mg(OEt)_2$ + $Al(Oi\text{-}Pr)_3$ + $Ti(Oi\text{-}Pr)_4$/$TiCl_4$	R_3Al	SL	Narrow to very broad	MWD control depending on catalyst mixture and type of cocat.; wider MWD obtained with a 2-steps polymn. process
Ger. Offen. 2, 115, 995 (1970)	Shell Int. Res.	$BuMgCl \cdot Bu_2O$ + $TiCl_4$	R_3Al	SL	Narrow to broad	MWD is broadened by catalyst precontacting with ethylene; MWD is narrowed by catalyst precontacting with hydrogen
Ger. Offen. 2, 139, 202 (1971)	Hoechst	"MT-Kontakte" prepared by contacting $Isoprenyl_3$ Al or Et_3Al with a mixture of the product (M) of reaction of $Isoprenyl_3Al$ and $TiCl_4$ with the product (T) of $Mg(OEt)_2$ + $TiCl_4$	R_3Al (especially $Isoprenyl_3Al$)	SL	MI_{15}/MI_5 7.8 to 12.6	MWD control with catalyst and cocat. composition; wider MWD obtained with a 2-steps polymn. process
U.S. 3, 899, 477 (1973)	Monsanto	$TiCl_4$ + $VOCl_3$ + $i\text{-}Bu_2AlH$	Me_3Al + Et_3Al_2-$(OEt)_3$	SL	Q 10.4 to 27.5	MWD can be controlled and broadened by varying the amount of $VOCl_3$; narrower MWD are obtained (Q 7.0) when Et_2AlCl is substituted for $Et_3Al_2(OEt)_3$
Ger. Offen. 2, 742, 585 (1976)	Asahi Chem. Ind.	$AlMg_6Et_{2.9}Bu_{12.1}$ + $SiHCl_3$/$TiCl_4$	$i\text{-}Bu_3Al$	SL	FR 68	FR 52 with Et_2AlCl_2 instead of $SiHCl_3$; control of MWD by changing catalyst and cocatalyst composition
Eur. Pat. Appl. 7 (1977)	Solvay	$Mg(OEt)_2$ + $Ti(OBu)_4$/ $AlRCl_2$	R'_3Al	SL	FR 30 to 49	MWD control by varying the nature of R and/or R'
Ger. Offen. 2, 809, 337 (1978)	Wacker Chemie	Methylhydrogenpolysiloxane + Et_2AlH/$TiCl_4$ + $Ti(OPr)_4$/$TiCl_3OPr$	R_3Al	SL	Broad	MWD control by varying catalyst composition and expecially the amount of $TiCl_3OPr$

U.S. 4, 154, 701 (1978)	Standard Oil	$Ti(OBu)_4$ + $Zr(OBu)_4$ + $VOCl_3$/$EtAlCl_2$/ deactivating agent	R_3Al	SL	Broad	With ROH or HCl as deactivating agent MWD is narroved or broadened respectively; MWD is wider for R = Et than for R = i-Bu
U.S. 4, 263, 422 (1979)	Dow Chem.	Dual catalyst system contg. an inorg. halide supported Ziegler cat. and an inorg. oxide supported Cr-contg. cat.	R_3Al	SO	Narrow to broad	MWD control con be obtained better with the dual cat. system than with either of the two catalysts alone
Ger. Offen. 3, 003, 327 (1979)	Sumitomo Chem.	$BuMgCl$ + Bu_2O + $SiCl_4$/$TiCl_3$/$TiCl_4$	$R_pY_q^-$, $Al(OR')_r$	SL	Broad	MWD control by varying the amount of $TiCl_3$ and the type of organo-aluminium compound; narrower MWD without $TiCl_3$
Ger. Offen. 2, 946, 562 (1979)	Asahi Chem. Ind.	(A) Organo-Mg compound + (B) Ti or V halide compound + (C) inorganic or organic Al, Si, Sn, Sb compound or (D) Ti or V compound	Organo-Al compound + (E) halogenated hydrocarbon	SL	Q 4.4 to 20.3	Polydispersity depending on catalyst and/or cocatalyst composition; broader MWD by using (C) and/or (D) and/or (E)

a. b) Conditions and foot notes see Table 10a

ty of this subject suggested us to treat it separately. While a lot of patent literature is available pertaining to the control of polypropylene stereoregularity, there is scarce, and not always significant informations about the possibilities of controlling its MWD [149-156]. On the other hand, it is particularly important, for the purposes of this review, to examine the relevant patent literature.

In Tables 10 to 13, the most significant patents dealing with the subject have been reported, beginning from 1968 (the year which can be said to have given rise to the industrial development of the "second generation" catalysts). This collection is not meant to be a comprehensive one, but contains the more relevant patents indicating the aim of the work. The selected patents have been classified according to four general methods of MWD control:

1) control based on the catalyst (Tables 10a–10d);
2) control based on the cocatalyst (Table 11);
3) Control based on the polymerization process (Table 12);
4) control based on a combination of the above mentioned methods (Table 13).

From this survey, significant tendencies can practically confirm what is theoretically predicted or represent empirical results of phenomena whose mechanisms are not yet completely known.

Polymers with broad MWD, suitable for application fields such as blow molding and extrusion, are obtained with catalysts prepared after proper choiche of:

a) the transition metal compound (sometimes a combination of more transition metal compounds);
b) the carrier;
c) the addition of a third compound;
d) other catalyst modification which are combinations of the mentioned a), b) and c) or which do not easily give rise to a generalized principle.

It is likely that the common feature of these methods of control is the alteration of the electronic environment of the active centre and the increase in the chemical physical heterogeneity of the catalytic surface.

A broadening of MWD can also be obtained, although less frequently, by modifying the composition of the cocatalyst (a high chlorine content in the aluminum alkyl seems sometimes to be effective).

A "multi-step" polymerization process and a simultaneous use of a proper catalytic system can give rise to a very broad MWD. This technique with many operative modifications, offers the possibility of regulating polyethylene MWD in a wide range. In fact by producing fixed amounts of polymers, according to determined molecular weights of each reactor and to the polymer final molecular weight, is possible to get the product with the desired MWD. This could be theoretically calculated by a simple equation [11], but one must also consider the residence time distributions in the various reactors.

Concerning the production of narrow MWD polyethylene, suitable for injection molding, the most commonly used method seems to be that based on catalyst chemical composition control (often by introducing electron-releasing ligands, i.e. alkoxides, on the transition metal compound). In this way minimum polydispersity values could be obtained (Q approximately 3 or even lower), a rather noteworthy result if one remembers that the catalytic systems dealt with are heterogeneous.

6 Conclusion

The MWD of polyolefins constitutes a property of fundamental importance, affects rheological and applicative properties of the polymers, and allows better understanding of Ziegler-Natta catalysis mechanism.

Although the study of polyolefins MWD is the object of everlasting attention, both from scientific and industrial research, there is no unified theory yet which explains the phenomena involved, enabling one to predict the MWD of a polymer with certainty. This is not surprising if one keeps in mind both the actual uncertainties of the experimental methods in MWD determination and the difficulties inherent to a rigorous kinetic study of Ziegler-Natta catalysis, deriving from the partial knowledge about the nature of the catalytic system, and by the difficulties encountered in trying to exactly identify macrophysical phenomena, which vary from monomer to monomer.

Among the various theories reported in the literature in an attempt to explain the observed polydispersity of polyolefins, which is generally broader than theoretically expected, those based only on the limitations due to mass transfer phenomena seem, for the most part, to be in contrast with several experimental evidences.

The chemical *theories* based on the *plurality* of active species present in the catalytic system, seem more convincing and more stimulating as a working hypothesis. Such *non-uniformity* of polymerization centres according to kinetic parameters should lead to MWD broadening, resulting from *superimposing* the individual distributions characterized, in the case of Ziegler-Natta polymerization, by a theoretical polydispersity index of about 2. Thus, at least in principle, it should be possible to regulate the MWD of the polymer by properly dosing type and concentration of active centres (homogeneous for narrow and heterogeneous for broad MWD).

Unfortunately, satisfactory analytical techniques are not yet available to precisely define number, chemical and structural composition of the active centres during polymerization. Thus, real capacity of MWD control is still a matter of more or less empirical approach.

However, from the examination of scientific and patent literature on polyethylene (unfortunately only few data are available on polypropylene), the following general principles for catalyst tayloring emerge:

— definitely soluble catalytic systems permit one to obtain very *narrow MWD*, even when the polymer is insoluble in the reaction medium;

— soluble catalytic systems, which undergo chemical and physical transformation during polymerization, and heterogeneous catalytic systems generally yield *broad MWD*, with polydispersity values varying over a wide range, especially for polyethylene;

— heterogeneous catalytic systems (the most interesting ones from an industrial point of view), characterized by a *plurality* of active species, which can be fully justified even from merely crystallographic considerations, can be adjusted by modifying the chemical and electronic environment of the transition metal. Thus, by varying the catalytic system ligands, the type of carrier and the possible Lewis base, the oxidation state of the transition metal is likely to be stabilized towards the action of the metal alkyl, thus leading to a greater active centre homogeneity.

This greater uniformity is reflected, not only in the narrowing of the MWD of the polymers, but also in the narrowing of the composition distribution of the copolymers.

A very broad MWD is generally obtained by operating with "multi-step" polymerization processes each of which yielding a polymer with different average molecular weight. The success of such an industrial arrangement however, always seems to be linked to the availability of suitable catalytic systems which supply intrinsically broad MWD (the patent literature often teaches the use of catalysts containing compounds of more than one transition metal).

Polymerization parameters, though constituting useful tool for studying polymerization mechanism, seem to be, in practice, less important than catalytic system for an effective control of MWD.

In the near future deeper investigations will be necessary in order to achieve an effective MWD control on the basis of theoretical knoweledge. These studies should be focused on the determination of number and structure of active centers, on elucidation of mechanism of reaction and kinetic constants relevant to the elementary stages; moreover, in propylene polymerization, some light could also be brought by a thorough understanding of the origin of regioselectivity and stereospecificity.

Acknowledgements: The authors wish to thank Montepolimeri S. p. A. for permission to publish this work. Acknowledgements are due to Prof. P. Galli, Director of Centro Ricerche G. Natta, Ferrara, for suggestions and helpful discussions. Special thanks are also due to Prof. Dr. P. Pino (ETH Zürich) for his encouragement and interest.

7 References

1. Galli, P., Luciani, L., Cecchin, G.: Angew. Makromol. Chem. *94*, 63 (1981)
2. Tait, P. J. T.: In Developments in Polymerization, R. N. Haward (ed.), Vol. 2, p. 81, London; Applied Science Publ., 1979
3. Chien, J. C. W. (ed.), Coordination Polymerization, New York; Academic Press, 1975
4. Tait, P. J. T.: In Macromolecular Chemistry, Vol. 1, p. 3, London; The Royal Society of Chemistry, 1980
5. Yermakov, Yu. I., Kuznetsov, B. N., Zakharov, V. A.: Catalysis by Supported Complexes, Amsterdam; Elsevièr, 1981
6. Boor, J.: Ziegler Natta Catalysts and Polymerization, New York; Academic Press, 1979
7. Berger, M. N., Boocock, G., Haward, R. N.: Adv. Catal. *19*, 24 (1969).
8. Reichert, K. H., Chem. Ing. Tech. *49*, 626 (1977)
9. Sivaram, S.: Ind. Eng. Chem., Prod. Res. Dev. *16*, (2), 121 (1977)
10. Cooper, W.: In Comprehensive Chemical Kinetics, C. H. Bamford, C. F. H. Tipper (eds.), Vol. 15, p. 189, Amsterdam; Elsevier, 1976
11. Böhm, L. L.: Angew. Makromol. Chem. *89*, 1 (1980)
12. Reich, L., Schindler, A.: In Polymerization by Organometallic Compounds, Polymer Review N. 12, p. 338, New York; Wiley, Interscience Publ., 1966
13. Keii, T.: Kinetics of Ziegler Natta Polymerization, London; Chapman and Hall, 1972
14. Vanderberg, E. J., Repka, B. C.: In Polymerization Processes, High Polymer Ser., 29, C. E. Schildknecht, I. Skeist (Eds.), p. 337, New York; Wiley Interscience Publ., 1977

15. Parker Forsman, J.: Hydrocarbon Processing 130, nov. 1972.
16. Sinn, H., Kaminsky, W.: Adv. Organomet. Chem. *18*, 99 (1980)
17. Sittig, M.: Polyolefin production Processes, Park Ridge, Noyes 1976
18. Foster, F. C.: Proceeding of the International Symposium on Macromolecules, Rio de Janeiro, E. B. Mano (Ed.), p. 209, Amsterdam; Elsevier, 1974
19. Short, J. N.: Ind. Res. Dev. 109, sept. 1980
20. Davis, T., Tobias, R. L., Peterli, E. B.: J. Polym. Sci. *56*, 485 (1962)
21. Japan. Patent 79 090, 291 (to Mitsui Toatsu Chem.)
22. Sakai, T.: Jpn. Steel Works, Tech. News *11*, 7 (1975)
23. Reichert, K. H.: Angew. Makromol. Chem. *94*, 1 (1981)
24. Sittig, M.: Catalysts and catalytic Processes 1967, Park Ridge, Noyes 1967
25. Doak, K. W., Schrage, A.: In ref. 26, p. 303
26. Raff, R. A. V., Doak K. W. (Eds.): Crystalline Olefin Polymers, Part I, High Polymers, vol. XX, New York; Wiley-Interscience, 1965
27. Boenig, H. W.: Polyolefins. Structure and Properties, New York; Elsevier, 1966
28. Stevens, J.: Hydrocarbon Processing. 179, nov. 1970
29. Martin, J. R., Johnson, J. F., Cooper, A. R.: J. Macromol. Sci.-Rev. Macromol. Chem. C *8*, 57 (1972)
30. Cottam, B. J.: J. Appl. Poym. Sci. *9*, 1853 (1965)
31. Saeda, S., Yotsuyanagi, J., Yamaguchi, K.: J. Appl. Polym. Sci. *15*, 277 (1971)
32. Platonov, M. P. et al.: Polym. Sci., U.S.S.R. *21*, 2640 (1980)
33. Braks, J. G., Huang, R. Y. M.: J. Polym. Sci., Polym. Physics Ed. *13*, 1063 (1975)
34. Mark, H. F., Gaylord, N. G., Bikales, N. M.: Encyclopedia of Polymer Science and Technology, New York; Wiley, Interscience vol. 7, p. 231 (1967) and vol. 9, p. 182 (1968)
35. Han, Chang Dae, Villamizar, C. A.: J. Appl. Polym. Sci., *22*, 1677 (1978)
36. Tung, L. H.: In ref. 26, p. 513
37. Nakajima, N.: Polym. Preprints, Division of Polymer Chem. A.C.S., *12*, (2), 804 (1971)
38. Servotte, A., Bruille, R. De: Makromol. Chem. *176*, 203 (1975)
39. Quackenbos, H. M.: J. Appl. Polym. Sci. *13*, 341 (1969)
40. Minoshima, W., White, J. L., Spruiell, J. E.: Polym. Eng. Sci. *20*, 1166 (1980)
41. Bersted, B. H., Slee, J. D., Richter, C. A.: J. Appl. Polym. Sci. *26*, 1001 (1981)
42. Bersted, B. H.: J. Appl. Polym. Sci. *19*, 2167 (1975)
43. Peebles, L. H.: Molecular Weight Distribution in Polymers, New York; J. Wiley-Interscience, 1971
44. Schulz, G. V.: Z. Phys. Chem. B *30*, 379 (1935)
45. Flory, P. J.: J. Am. Chem. Soc. *58*, 1877 (1936)
46. Matkovskii, R. Ye., Irzhak, V. I., D'yachkovskii, F. S.: Vysokomol. Soyed. A *19*, (9), 2073 (1977)
47. Henrici-Olivé, G., Olivé, S.: Adv. Polym. Sci. *15*, 1 (1974)
48. Henrici-Olivé, G., Olivé, S.: In ref. 3, p. 291
49. Henrici-Olivé, G., Olivé, S.: Angew. Chem. 82, 255 (1970)
50. Henrici-Olivé, G., Olivé, S.: Angew. Chem. Int. Ed. Engl. *9*, 243 (1970)
51. Longi, P., Greco, F., Rossi, U.: Chim. Ind. (Milano) *55*, 253 (1973)
52. Langer, A. W.: J. Macromol. Sci. Chem. A *4*, 775 (1970)
53. Attridge, C. J. et al.: Chem. Comm. 132 (1973)
54. Agasaryan, A. B., Belov, G. P., Davtyan, S. P., Eritsyan, M. L.: Europ. Polym. J. *11*, 549 (1975)
55. Böhm, L. L., Passing, H.: Makromol. Chem. *177*, 1097 (1976)
56. Grieveson, B. M.: Makromol. Chem. *84*, 93 (1965)
57. Meyer. K., Reichert, K. H.: Angew. Makromol. Chem. *12*, 175 (1970)
58. Reichert, K. H.: Angew. Makromol. Chem. *13*, 177 (1970)
59. Schnell, D., Fink, G.: Angew. Makromol. Chem. *39*, 131 (1974)
60. Fink, G. et al.: J. Appl. Polym. Sci. *20*, 2779 (1976)
61. Francis, P. S., Cook, R. C., Elliott, J. H.: J. Polym. Sci. *31*, 453 (1958)
62. Henry, P. M.: J. Polym. Sci. *36*, 3 (1959)
63. Kenyon, A. S., Salyer, I. O.: J. Polym. Sci. *43*, 427 (1960)
64. Tung, L. H.: J. Polym. Sci., *24*, 333 (1957)
65. Wesslau, H.: Makromol. Chem. *26*, 102 (1958)

66. Wesslau, H.: Makromol. Chem. *20*, 111 (1956)
67. Wijga, P. W. O., Van Schooten, J., Boerman, J.: Makromol. Chem. *36*, 115 (1960)
68. Pegoraro, M.: Chim. Ind. (Milano) *44*, 18 (1962)
69. Tanaka, S., Nakamura, A., Morikawa, H.: Makromol. Chem. *85*, 164 (1965)
70. Davis, T. E., Tobias, R. L.: J. Polym. Sci. *50*, 227 (1961)
71. Hirooka, M., Kanda, H., Nakaguchi, K.: J. Polym. Sci. Part B *1*, 701 (1963)
72. Westerman, L.: J. Polym. Sci. Part A *1*, 411 (1963)
73. Chiang, R.: J. Polym. Sci. *28*, 235 (1958)
74. Shyluk, S.: J. Polym. Sci. *62*, 317 (1962)
75. Mendelson, R. A.: J. Polym. Sci. Part A *1*, 2361 (1963)
76. Quano, A. C., Mercier, P. L.: J. Polym. Sci. Part C *21*, 309 (1968)
77. Yamaguchi, K.: Makromol. Chem. *132*, 143 (1970)
78. Williamson, G. R., Wright, B., Haward, R. N.: J. Appl. Chem. *14*, 131 (1964)
79. Tung, L. H.: J. Polym. Sci. *20*, 495 (1956)
80. Pechočová, M., Pechoč, V.: Makromol. Chem. *163*, 235 (1973)
81. Natta, G. et al.: Chim. Ind. (Milano) *44*, 621 (1962)
82. Boor, J.: Macromol. Rev. *2*, 115 (1967)
83. Clark, A., Bailey, G. C.: J. Catal. *2*, 230 (1963)
84. Clark, A., Bailey, G. C.: J. Catal. *2*, 241 (1963)
85. Natta, G.: J. Polym. Sci. *34*, 21 (1959)
86. Gordon, M., Roe, R. J.: Polymer *2*, 41 (1961)
87. Natta, G., Pasquon, I., Zambelli, A.: J. Am. Chem. Soc. *84*, 1488 (1962)
88. Zambelli, A., Natta, G., Pasquon, I.: J. Polym. Sci. Part C *4*, 411 (1964)
89. Bogdanović, B., Spliethoff, B., Wilke, G.: Angew. Chem. In. Ed. Engl. *19*, 622 (1980)
90. Giannini, U., Zucchini, U., Albizzati, E.: Polym. Sci. Polym. Lett. Ed. *8*, 405 (1970)
91. Henrici-Olivé, G., Olivé, S.: Adv. Polym. Sci. *6*, 421 (1969)
92. Henrici-Olivé, G., Olivé, S.: Coordination and Catalysis, New York; Verlag Chemie, 1977
93. Ballard, D. G. H.: Adv. Catal. *23*, 263 (1973)
94. Henrici-Olivé, G., Olivé, S.: Adv. Polym. Sci. *15*, 1 (1974)
95. Breslow, D. S., Newburg, N. R.: J. Am. Chem. Soc. *81*, 81 (1959)
96. Carrick, W. L. et al.: J. Am. Chem. Soc. *82*, 3883 (1960)
97. Christman, D. L.: J. Polym. Sci. Part A-1 *10*, 471 (1972)
98. Cozewith, C., Ver Strate, G.: Macromolecules *4*, 482 (1971)
99. Soga, K., Ikeda, S., Keii, T.: Makromol. Chem. *178*, 337 (1977)
100. Dumas, P., Martineau, D., Sigwalt, P.: IUPAC, Macro Florence 1980. Intern. Symp. on Macromolecules, Preprints, vol. 2, p. 71
101. Belov, G. P. et al.: Izv. Akad. Nauk U.S.S.R., Ser. Chem. 1275 (1966)
102. Belov, G. P. et al.: Vysokomol. Soedin. A *9*, N° 6, 1269 (1967)
103. Belov, G. P. et al.: Europ. Polym. J. *6*, 29 (1970)
104. Belov, G. P. et al.: Makromol. Chem. *140*, 213 (1970)
105. Waters, J. A., Mortimer, G. A.: J. Polym. Sci. Part A-1 *10*, 1827 (1972)
106. Waters, J. A., Mortimer, G. A.: J. Polym. Sci. Part A-1 *10*, 895 (1972)
107. Höcker, H., Saeki, K.: Makromol. Chem. *148*, 107 (1971)
108. Doi Y. et al.: J. Polym. Sci. Polym. Chem. Ed. *13*, 2491 (1975)
109. Keii, T.: Makromol. Chem. *180*, 57 (1979)
110. Doi, Y., Ueki, S., Keii, T.: Makromol. Chem. *180*, 1359 (1979)
111. Doi, Y., Ueki, S., Keii, T.: Macromolecules *12*, 814 (1979)
112. Doi, Y., Ueki, S., Keii, T.: Polymer *21*, 1352 (1980)
113. Ueki, S., Doi, Y., Keii, T.: Makromol. Chem., Rapid Commun. *2*, 403 (1981)
114. Rasmussen, D. M.: Chem. Eng. 104, (Sept. 18, 1972)
115. Bateman, H. L.: Plastics Eng. *31*, 73 (April 1974)
116. Mussa, C., I. V.: J. Polym. Sci. *26*, 76 (1957)
117. Mortimer, G. A.: J. Appl. Polym. Sci. *20*, 55 (1976)
118. Schindler, A.: Makromol. Chem. *118*, 1 (1968)
119. Schindler, A.: Makromol. Chem. *114*, 77 (1968)
120. Henrici-Olivé, G., Olivé, S.: Chem. Ing. Techn. *43*, 906 (1971)
121. Nowakowska, M. et al.: Polym. Sci. U.S.S.R. *20*, 2523 (1979)
122. Kühlein, K., Clauss, K.: Makromol. Chem. *155*, 145 (1972)

123. Sangalov, Yu., Il'yasova, A. I., Minsker, K. S.: Polym. Sci. U.S.S.R. *18*, 876 (1976)
124. Tajima, Y., Kunioka, E.: J. Polym. Sci. Part A-1 *6*, 241 (1968)
125. Schreyer R. C. (to E. I. du Pont de Nemours): U.S.2.986.531 (1968)
126. Herrmann, Ch., Streck, R.: Angew. Makromol. Chem. *94*, 91 (1981)
127. See ref. 6, Chapter 9
128. Erofeev, B. V. et al.: Vestsi Akad. Navuk Belarus. S.S.R., Ser. Khim. Navuk. *5* (1971) (3)
129. Kircheva, R. S. et al.: Plast. Massy *7*, 3 (1971)
130. Combs, R. L. et al.: J. Polym. Sci. Part A-1 *5*, 215 (1967)
131. Doi, Y., Nishimura, Y., Keii, T.: Polymer *22*, 469 (1981)
132. Mitsubishi: Chem. Ind., Eur. Pat. Appl. 0, 039, 442 (1980)
133. Ivanchev, S. S., Baulin, A. A., Rodionov, A. G.: J. Polym. Sci., Polym. Chem. Ed. *18*, 2045 (1980)
134. Zakharov, V. A., Bukatova, Z. K., Makhtarulin, S. I., Chumayevskii, N. B., Yermakov, Yu. I.: Polym. Sci. U.S.S.R. *21*, 544 (1979)
135. Baulin, A. A.: Polym. Sci. U.S.S.R. *22*, 205 (1980)
136. Petrova, T.: Europ. Polym. J. *12*, 571 (1976)
137. Lassalle, D., Vidal, J. L., Roustant, J. C., Mangin, P.: Paper at the 5th Europ. Plastics and Rubbers Conference, Paris 1978
138. Carrick, W. L., Turbett, R. J., Karol, F. J., Karapinka, G. L., Fox, A. S., Johnson, R. N.: J. Polym. Sci., Part A-1 *10*, 2609 (1972)
139. Böhm, L. L.: Polymer *19*, 562 (1978)
140. Böhm, L. L.: Polymer *19*, 553 (1978)
141. Greco, A., Bertolini, G., Cesca, S.: J. Appl. Polym. Sci. *25*, 2045 (1980)
142. Suzuki, E. et al.: Makromol. Chem. *180*, 2235 (1979)
143. Suzuki, E. et al.: IUPAC, Macro Florence 1980, Intern. Symp. on Macromolecules, Preprints, vol. 2, p. 52
144. Keii, T.: Paper at the MMI Intern. Symp. on Transition Metal Catalyzed Polymerization: Unsolved Problems, Midland, August 1981
145. Doi, Y., Suzuki, E., Keii, T.: Makromol. Chem. Papid Commun. *2*, 293 (1981)
146. Montedison: Ger. Offen. 2, 230, 672 (1971); Montedison and Mitsui Petrochem. Ind., Ger. Offen. 2, 643, 143 (1975) and 2, 735, 672 (1976)
147. Soga, K., Terano, M., Ikeda, S.: Polymer Bull. *1*, 849 (1979)
148. Kashiwa, N.: Paper presented at the MMI Intern. Symp. on Transition Metal Catalyzed Polymerization: Unsolved Problems, Midland, August 1981
149. Standard Oil, U.S. 4, 276, 191 (1979)
150. Sumitomo Chem., Eur. Pat. Appl. 0, 021, 753 (1979)
151. Showa Denko, Jpn. Kokai Tokkyo Koho 79 85, 289 (1977)
152. Chisso, Ger. Offen. 2, 920, 799 (1978)
153. Chisso, Ger. Offen. 2, 922, 751 (1978)
154. Chisso, It. Pat. Appl. 48080 A/81 (1980)
155. El Paso: U.S. 4, 284, 738 (1980)
156. El Paso, Fr. Demande 2, 483, 429 (1980)
157. See ref. 6, Chapter 12
158. Gaylord, N. G., Mark, H. F.: Linear and Stereoregular Addition Polymer, Interscience, New York, p. 132, 1959
159. Baulin, A. A., Shalaeva, L. F., Ivanchev, S. S.: Dokl. Akad, Nauk SSSR, *231* (2), 413 (1976); C.A. *86*, 44096 r (1977)
160. Severova, N. N. et al.: Plast. Massy *24* (1976); C.A. *86*, 73187b (1977)
161. Cihlář, J. et al.: Makromol. Chem. *182*, 1127 (1981)
162. Cihlář, J. et al.: Makromol. Chem. *181*, 2549 (1980)
163. Cihlář, J., Meijzlik, J., Hamřík, O.: Makromol. Chem. *179*, 2553 (1978)
164. Chien, J. C. W.: J. Polym. Sci., Part A *1*, 1839 (1963)
165. Chien, J. C. W.: J. Am. Chem. Soc. *81*, 86 (1959)
166. Noristi, L.: Personal communication
167. Taylor, W. C., Tung, L. H.: Polymer Letters *1*, 157 (1963)
168. Meyer, H., Reichert, K. H.: Angew. Makromol. Chem. *57*, 211 (1977)
169. Crabtree, J. R. et al.: J. Appl. Polym. Sci. *17*, 959 (1973)
170. Schnecko, H., Jung, K. A., Kern W.: In (3), p. 73

171. Jung, K. A., Schnecko, H.: Makromol. Chem. *154*, 227 (1972)
172. Bier, G., Gumboldt, A., Lehmann, G.: Plast. Inst., Trans. J. *28*, 98 (1960)
173. Bier, G.: Makromol. Chem. *70*, 44 (1964)
174. Bier G. et al.: Makromol. Chem. *58*, 1 (1962)
175. Doi, Y., Keii, T.: Makromol. Chem. *179*, 2117 (1978)
176. Rishina, L. A., Vizen, E. I.: Europ. Polym. J. *16*, 965 (1980)
177. Doi, Y., Morinaga, A., Keii, T.: Makromol. Chem. Rapid Commun. *1*, 193 (1980)
178. Böhm, L. L.: Polymer *19*, 545 (1978)
179. Combs, R. L., Slonaker, D. F., Coover, H. W.: Jr., J. Appl. Polym. Sci. *13*, 519 (1969)
180. Pasquon, I., Dente, M., Narduzzi, F.: Chim. Ind. (Milano) *41*, 387 (1959)
181. Coover, H. W.: Jr., J. Polym. Sci., Part C C *4*, 1511 (1964)
182. Buls, V. W., Higgins, T. L.: J. Polym. Sci., Part A-1 *8*, 1025 (1970)
183. Buls, V. W., Higgins, T. L.: J. Polym. Sci, Part A-1 *8*, 1037 (1970)
184. Nagel, E. J., Kirillov, W. A., Harmon, Ray W.: Ind. Eng. Chem. Prod. Res. Dev. *19*, 372 (1980)
185. Taylor, T. W., Choi, K. Y., Yuan, H., Harmon Ray, W.: Paper at the MMI Intern. Symp. on Transition Metal Catalyzed Polymerization: Unsolved Problems, Midland, August 1981
186. Schmeal, W. R., Street, I. R.: AIChE J. *17*, 1188 (1971)
187. Singh, D., Merril, R. P.: Macromolecules *4*, 599 (1971)
188. Hock, C. W.: J. Polym. Sci., Polym. Chem. Ed. *4*, 3055 (1966)
189. Vizen, Ye. I., Yakobson, F. I.: Polym. Sci. U.S.S.R. *20*, 1046 (1979)
190. Galli, P., Noristi, L.: Paper presented at the 5th Europ. Plastics and Rubbers Conference, Paris 1978
191. Chien, J. C. W.: J. Polym. Sci., Polym. Chem. Ed. *17*, 2555 (1979)
192. See ref. 6, Chapter 8, p. 190
193. See ref. 6, Chapter 8, p. 180
194. Feldman, C. F., Perry, E.: J. Polym. Sci. *46*, 217 (1960)
195. Bier, G., Gumboldt, A., Schleitzer, G.: Makromol. Chem. *58*, 43 (1962)
196. Hoffman, A. S., Fries, B. A., Condit, P. C.: J. Polym. Sci., Part C *4*, 109 (1964)
197. Fisa, B., Marchessault, R. H.: J. Polym. Sci., Polym. Letters Ed. *12*, 561 (1974)
198. Unpublished results of our laboratories
199. Buls, V. W., Higgins, T. L.: J. Polym. Sci., Polym. Chem. Ed. *11*, 925 (1973)
200. Muñoz-Escalona, A., Villalba, J.: Polymer *18*, 179 (1977)
201. Mussa, C., I. V.: J. Appl. Polym. Sci. *1*, 300 (1959)
202. Roe, R. J.: Polymer *2*, 60 (1961)
203. Schindler, A.: Monatsch. Chem. *95*, 868 (1964)
204. Boor, J., Jr.: Ind. Eng. Chem. Prod. Res. Develop. *9*, 437 (1970)
205. McGreavy, C.: J. Appl. Chem. Biotechnol. *22*, 747 (1972)
206. Zakharov, V. A. et al.: Makromol. Chem. *178*, 967 (1977)
207. Tait, P. J. T.: Chem. Tech. *5*, 688 (1975)
208. See ref. 3, p. 155
209. Zucchini, U.: Unpublished results
210. Hoff, R. E., Shida, M.: J. Appl. Polym. Sci. *17*, 3003 (1973)
211. Denbigh, K.: Chemical Reactor Theory, Cambridge University Press, 1965
212. Biesenberger, J. A., Tadmor, Z.: Polymer Eng. Sci. 299, october 1966
213. Zak, A. V., Shpakov, P. P.: Zh. Prikl. Khim. (Leningrad) *54* (4), 932 (1981)
214. Chappelar, D. C., Simon, R. H.: Addition and Condensation Polymerization Processes, Advances in Chem. Ser. *91*, R. F. Gould (Ed.), Am. Chem. Soc. Publ., Washington, D.C., 1969, p. 1
215. Allegra, G.: Makromol. Chem. *145*, 235 (1971)
216. Corradini, P. et al.: Europ. Polym. J. *15*, 1133 (1979)
217. Corradini, P. et al.: Europ. Polym. J. *16*, 835 (1980)
218. Bukatov, G. D. et al.: Makromol. Chem. *179*, 2093 (1978)
219. Minsker, K. S. et al.: Polym. Sci. U.S.S.R. *22*, 2478 (1980)
220. Ponomarev, O. A. et al.: Vysokomol. Soyed. A *17*, 309 (1975)
221. Kohara, T. et al.: Makromol. Chem. *180*, 2139 (1979)
222. Cossee, P.: Tetrahedron Lett. *17*, 12 (1960)
223. Cossee, P.: J. Catal. *3*, 80 (1964)

224. Cossee, P.: Rec. Trav. Chim. Pays-Bas *85*, N°. 9–10, 1152 (1966)
225. Cecchin, G.: Unpublished results
226. Montedison, U.S. 4.298.718 (1968); G.B. 1.292.853 (1968); G.B. 1.305.610 (1969)
227. Giannini, U.: Makromol. Chem., Suppl. *5*, 216 (1981)
228. Nielsen, R. P.: Paper at the MMI Intern. Symp. on Transition Metal Catalyzed Polymerization: Unsolved Problems, Midland, August 1981
229. Karol, F. J., Carrick, W. L.: J. Am. Chem. Soc. *83*, 2654 (1961)
230. Baulin, A. A. et al.: Plast. Massy 9 (9) (1980)
231. Vizen, Ye., Kissin, Yu. V.: Polym. Sci. U.S.S.R. *11*, 2017 (1969)
232. Arlman, E. J., Cossee, P.: J. Catal. *3*, 99 (1964)
233. Thiele, E. W.: Ind. Eng. Chem. *31*, 916 (1939)
234. Erenburg, Ye. G.: Polym. Sci. U.S.S.R. *22*, 1841 (1980)
235. Doi, Y. et al.: Polymer *23*, 258 (1982)
236. Rodionov, A. G. et al.: Vysokomol, Soedin., Ser. A *23*, 1560 (1981)
237. Böhm, L. L.: Makromol. Chem. *182*, 3291 (1981)
238. Kashiwa, N., Yoshitake, J.: Makromol. Chem. Rapid Commun. *3*, 211 (1982)
239. Graff, R. J. L., Kortleve, G., Vonk, C. G.: Polymer Letters *8*, 735 (1970)
240. For ethylene-propylene-based terpolymers see: S. Cesca, Macromol. Rev. *10*, 1 (1975)
241. See ref. 240, p. 33
242. Yuan, H. G., Taylor, T. W., Choi, K. Y., Ray, W. H.: J. Appl. Polym. Sci. *27*, 1691 (1982)
243. Lee, C. K., Bailey, J. E.: AIChE J. *20* (1), 74 (1974)
244. Gol'denberg, A. L. et al.: Vestsi Akad. Nauk Belarus. SSR, Ser. Khim. Nauk (1) 29 (1973)
245. Rishina, L. A., Pirogov, O. N., Chirkov, N. M.: Polym. Sci. U.S.S.R. *11*, 1124 (1969)
246. Keii, T., Doi, Y., Kobayashi, H.: J. Polym. Sci., Polym. Chem. Ed. *11*, 1881 (1973)
247. Intermediate situations have been theoretically predicted under particular conditions; see: G. Natta, I. Pasquon, M. Dente, Chim. Ind. (Milano) *44*, 1 (1962); A. Guyot, J. Polym. Sci., Part B *6*, 123 (1968)
248. Montedison, U.S. 4.124.532 (1975)
249. Nunes, R. W., Martin, J. R., Johnson, J. F.: Polym. Eng. Sci. *22*, 205 (1982)
250. Pino, P., Mülhaupt, R.: Angew. Chem. Int. Ed. Engl. *19*, 857 (1980)
251. Fink, G., Schnell, D.: Angew. Makromol. Chem. *105*, 15 (1982); ibid. 31; ibid. 39
252. Corradini, P., Guerra, G., Barone, V.: Preprints IUPAC 28th Macromol. Symposium, Amherst, Ma., July 12–16, p. 235, 1982
253. Keii, T. et al.: Preprints IUPAC 28th Macromol. Symposium, Amherst, Ma., July 12–16, p. 237, 1982
254. Choi, K. Y., Taylor, T. W., Ray, W. H.: Preprints IUPAC 28th Macromol. Symposium, Amherst, Ma., July 12–16, p. 240, 1982
255. Schulz, G. V.: Z. Physik, Chem. *47*, 155 (1940)
256. Huggins, M. L., Okamoto, H.: Polymer Fractionation, New York; Academic Press, 1966
257. Dawkins, J. V.: Development in Polymer Characterization, vol. 1, London; Appl. Sci. Publ., 1978
258. Zabusky, H. H., Heitmiller, R. F.: SPE, Trans. *4* (1), 17 (1964)

Received November 22, 1982
H.-J. Cantow (editor)

Foamed Polymers. Cellular Structure and Properties

Fyodor A. Shutov

Physical Department, Building Institute, Leningrad 198005, USSR

Nichts ist drinnen, nichts ist draussen:
Denn was innen, das ist außen.
J. W. v. Goethe

This survey deals with the fundamental morphological parameters of foamed polymers including size, shape and number of cells, closeness of cells, cellular structure anisotropy, cell size distribution, surface area etc. The methods of measurement and calculation of these parameters are discussed. Attempts are made to evaluate the effect and the contribution of each of these parameters to the main physical properties of foamed polymers namely apparent density, strength and thermoconductivity. The cellular structure of foamed polymers is considered as a particular case of porous statistical systems. Future trends and tasks in the study of the morphology and cellular structure-properties relations are discussed.

Abbreviations

FL-1 Soviet grade resol phenol-formaldehyde foam
FRP-1 Soviet grade resol-formaldehyde foam
GSE Gas-Structural Element
PPU-3 Soviet grade rigid polyurethane foam
PSB, PSB-S, PS-4 Soviet grades polystyrene foam
PVKH-1 Soviet grade rigid poly(vinyl chloride) foam

Physicochemical Symbols

A	normalization constant
b	side of cube
B	number of angles
C	constant
d	cell diameter
d_{cr}	critical (minimum) cell diameter
d_{max}	maximum cell diameter
D	cell diameter, width of cell
D_a	arithmetical cell diameter
D_g	geometrical cell diameter
D_i	current cell diameter
D_n	nominal cell diameter
D_s	surface cell diameter
D_p	probable cell diameter
D_v	volume cell diameter
$D^{\parallel}, D^{\perp}$	cell diameters parallel and perpendicular to heat flux
E_f	flexibility modulus
E_p	modulus of polymer phase
E_0	modulus of rod
g	weight of polymer phase
g_c	weight of polymer per cell
G	gas filling
G_f	gas-filling factor
G_m	active gas filling
h	height of cell
k	length ratio; parameter
K_0	Hankel function
l	intermediate layer depth
m	mass of dry foam
m_g	mass of gas in foam
m_w	mass of wet foam
M_p	mass of polymer phase of foam
N	total number of cells (spheres)
N_b	number of cells in a cubic
N_i	current number of cells

N_V number of cells having a volume V
p   parameter
P   probability; load
q, q_1, q_2, q_3 anisotropy coefficients
q^* effective anisotropy coefficient
Q   weight of foam
r   cell (spherical) radius
r_c curvature radius
r_h hydraulic radius
r_{max} maximum radius
r_p planar projection radius
R_B roughness factor
R_c inscribed circle radius
S   surface area
ΔS surface area increment
S_a cross-sectional area
S_c surface area of cell
S_d surface area of conducting cells
S_{ext} external surface area
S_{int} internal surface area
S_s surfase area of spherical cell
S_{sp} specific surface area
S_{sp}^K kinetic specific surface area
S_{sp}^M mass specific surface area
S_{sp}^V volumetric specific surface area
S_{vis} visible surface area
S_Σ total surface area
S_0 original cross-sectional area
ΔT temperature gradient
U   filling ratio
v, v_1, v_2, v_3 wave propagation velocities
v_g wave propagation velocity in gas phase
v_p wave propagation velocity in polymer phase
V   packed volume
V_c cell volume
V_d volume of dry foam
V_{eff} effective cell volume
V_g volume of gas phase
V_m volume of cells not participating in filtration process
V_{max} volume containing a maximum number of cells
V_s volume òf solid (polymer) phase
V_{sp} specific cell volume
V_0 net volume of foam
W   sphere size distribution function
x   depth of cell; distance between spheres
X   section plan
Z   transparency

α	angle; parameter
α_s	surface aspect ratio
α_v	volume aspect ratio
β	width-to-length ratio
γ	apparent density (volumetric weight) of foam
γ_p	density of polymer phase; density of unexpanded polymer
γ_m	minimum of γ
γ_y	value of γ at distance "y"
γ_1	increment of γ
γ^*	true apparent density of foam
$\Delta\gamma$	gradient of γ
Γ	gamma function
δ	wall thickness of cell
δ_{cr}	critical (minimum) wall thickness of cell
ε	deformation; strain
ε_c	compressive deformation
ε_{max}	maximum deformation
ε_t	tensile deformation
η	distribution median
ϑ_g	volume fraction of gas phase
ϑ_p	volume fraction of polymer phase
ϑ_p^c	value ϑ_p for closest packed foam
ϑ_w	volume fraction of water phase
ϑ_w^c	value ϑ_w for closest packed foam
ϑ_α	volume fraction of open cells
ϑ_α^c	value ϑ_α for closest packed foam
$\varkappa$	coefficient of variation
λ	thermoconductivity
λ_c	convective thermoconductivity
λ_g	thermoconductivity of gas phase
λ_r	radiative thermoconductivity
λ_s	thermoconductivity of solid phase
μ_i	D_i/D
ν	number of contacts
ν	N_i/N
ξ	G/U
σ	standard deviation; root-mean-square deviation
σ_c	compressive strength
σ_t	tensile strength
φ	roundness of cell shape
ψ	sphericity of cell shape
ω	configuration factor

1 Introduction

At present, there are at least two approaches to the investigation of the cellular structure of foamed polymers. In the first one, which may formally be called a graphical approach, attempts are made to draw conclusions on the macroscopic properties of foamed polymers from morphological parameters such as the geometry and stereometry of cells of various sizes, shapes and types. The second approach, which may be referred to as physicochemical, attempts to explain and predict polymer morphology from the data on the chemical composition of the polymer matrix and the mechanisms of foaming [1].

Generally, a physico-chemical treatment of the problem requires the accumulation of descriptive graphical data; therefore, most efforts have been concentrated on the first approach for many years, resulting in the compilation of a wealth of experimental data which now urgently needs systematization.

The aim of this review is 1) systematizing the graphical data and 2) attempting, on the basis of such systematization, first to outline and then to formulate clearly the basic relationships between the morphology and the properties of foamed polymers.

2 Porous and Cellular Systems

2.1 Statistical Meaning of "Pore" and "Cell" Concepts

For a few decades now cellular and porous systems have been classified in morphological terms by simulating the real systems by one or another imaginary, and always simplified, geometrical or stereometrical scheme using an artificially ordered-structure model. Such classifications have always been based on the concept that in any cellular or porous system it is possible to isolate a structural element (cell or pore). However, the diversity of pore and cell types even in small-sized real foamed systems does, in most cases, not permit a definition by only one single geometrical structural parameter, as for other types of solids (type and volume of elementary cell, interplanar or interatomic distances, etc.) [1].

Since it is difficult and sometimes even impossible [2] to give a clear-cut definition of the terms "pore" and "cell" with respect to real systems, agree with the now widely accepted opinion of Scheidegger that a rigorous and correct characterization of porous media may be only obtained by statistical methods [3-7].

Today, it is becoming evident that the former classifications of cellular or foamed plastics based on such morphological notion as a "cell" or "pore" and the type of communication (degree of closeness) between individual cells and pores needs be modified. An analysis of the information obtained in recent years reveals that it is not only the size and shape of the cell itself which is the basic morphological unit determining the properties of gas-filled polymers but also the size and configuration of the intercellular space filled with the polymer matrix, i.e. the walls and struts of cells. Indeed, if the size and shape of cells and the proportion of communicating cells are the same, for many foamed polymers considerable differences in the principal physico-mechanical properties, such as elasticity and strength, can be attributed either to the different shapes and sizes of the struts or the differences in cellular wall thickness.

We therefore believe that the taditional "cell" concept should now be replaced by a new morphological notion, the gas structural element (GSE) [8]. The GSE is a statistically averaged model of a spatial structure consisting of a gas cavity (a cell) and its walls and struts. It is the elementary gas and solid phase unit which recurs within a spezific period and at a high degree of ordering in either a part or the total volume, constituting the foamed plastic macrostructure.

The gas structural element concept is more general than the "cell" concept. Indeed, similarly sized and shaped cells may form different types of GSE (due to different configurations and different intercellular space volume). On the other hand, a similarity of the GSE would indicate similar size and shapes of the cells as well as their walls, struts and the intercellular space (see Sect. 5.3).

Among the structural models of cellular or porous materials those characterized by differently ordered packing of balls or spheres of the same diameters have most widely been used. In this approach either the spheres have been considered as real cells or the cell (pore) models have been derived from an analysis of assumed spacings between the contacting solid spheres. However, no system of packed spheres would adequately describe the properties of any real cellular system which never exhibit a regular packing. It is also impossible to describe the structure of most cellular systems via models assuming spheres of equal size.

2.2 Growth Systems and Addition Systems

It is still impossible to define any single classification criterion for the diversity of real porous and cellular systems. However, Radushkevich [2] believes that there are at least two criteria useful for the classification. These criteria include the mechanism of formation (the genesis) of a system and the general nature of the structure. According to the first criterion all systems may be divided into two major groups, commonly referred to as growth and addition systems. As regards the general structural differences, they can be divided into systems with well-ordered and unordered structure. The above two-component classification allows most cellular and porous materials known at present to be classified, there are more complex and frequently occurring systems which may be regarded as a combination of growth and addition systems.

Such two-component classification is sufficient for the purpose in hand since it covers all the known types of formed plastics.

A porous system resulting from dispersion of a (macroscopically) continuous medium, condensation, a chemical reaction, or from any other specific process (e.g. physical or biological) may be called a *growth system*. Such a system usually possesses an inimitable morphology. Growth systems include the following natural or man-made porous materials: pumice, cokes, activated carbon, carbon, ceolites, cellulose fibers, and finally most foamed polymers.

If, on the other hand, a system is produced by a random conglomeration of a large number of individual elements (porous or nonporous), the result will be an *addition system*. They also include both natural and man-made materials: sand, gravel, woven fabrics, powders, fibrous materials, paper. Addition systems also include a special type of gas-filled plastics, the so-called syntactic foams.

We designate as composite (combined) such systems as are formed by combination of growth and addition systems. As a rule, such systems are produced both by the addition of individual elements and through growth of the individual elements (e.g. pore formation or foaming). Composite systems are, for exemple, ceramics, foam glass, fabrics, Gouch filters, membrane filters, most construction materials, metallic or polymeric cakes, etc.

Plastics manufactured from prefoamed (pre-expanded) granules such as polystyrene are also composite systems according to the proposed classification.

Whether or not it is possible to describe the structure of a growth system depends directly on the kinetics of the processes according to which the systems are formed. In the case of foamed polymers such processes include liberation of gas, gas bubble formation, growth and stabilization, temperature variations, the manner in which the formulation is prepared, etc. Unfortunately, there are few informations available on the morphology of such systems. Therefore, it is not always possible to analyze the characteristics of a polydisperse structure formed in a growth system. Such structures are either strictly regular (macroscopically regular) or pseudoregular (spontaneous). Conversely, addition systems are not formed by a specific process and the elements (particles) of a given system are arranged in a random manner. In a simplest version they may be regarded as a bulk conglomeration of individual elements of some shape.

3 Gas-Filling of Foamed Polymers

3.1 Terminology

In general, gas-filled systems are characterized by the porosity which is the simplest statistical parameter of the majority of the systems studied. However, in foamed polymers the term "porosity" should be replaced by the term "gas-filling" and the term "porosity factor" by "gas-filling factor", respectively, taking into account the fact that interactions between the polymeric and the gaseous phase occur. In contrast to mineral porous systems (foamed glass, foamed ceramics, foamed concrete, etc.), the systems are affected by the gas phase composition, i.e. by the nature of the gas filling the voids, thereby "creating" porosity. Even with the same expansion process at the same composition of the monomer mixture, the porosity factor (volumetric weight) of the resultant foamed plastic may be varied within very broad limits simply by modifying the chemical composition of the blowing gas. Foamed plastics also differ from mineral porous materials since the chemical nature of the blowing gas and its pressue in the voids (cells) affect the behavior and physical properties of the foamed polymers, not only immediately on their formation but also over long periods of their use.

3.2 Basic Relationships

We will now consider gas-filled polymeric systems as bodies composed of a very large number of individual particles containing spaces. In general these spaces may be either void or filled with gas or liquid. For simplicity, the particles themselves will be assumed to be non-porous.

Let a sufficiently portion of the porous system have a net volume of V_0 which is composed of the volume of the substance (the solid phase) V_s and the total volume of all voids V_g. The gas filling G will then be given by the ratio of V_g to the total of the system:

$$G = \frac{g}{V_g + V_s} = \frac{g}{V_0} \tag{1}$$

The parameter G represents the net (or total) gas filling of the structural elements, GSE. In order to account for the portion of the GSE which, under specific conditions, can take part in liquid or gas filtration, we have to introduce a concept of "active gas-filling" (or "active porosity") G_m:

$$G_m = \frac{V_g - V_m}{V_0} \tag{2}$$

where V_m is the volume of those GSE which are not involved in the filtration process.

It is often convenient to use the "gas-filling factor" G_f in place of the term "gas-filling":

$$G_f = \frac{V_g}{V_s} \tag{3}$$

The values G and G_f are related by:

$$\frac{1}{G} = \frac{1}{G_f} + 1$$

If we neglect the weight of the gas inside the foamed plastic the volume fraction ϑ_g of cells (pores) will be:

$$\vartheta_g = \frac{\gamma_p - \gamma}{\gamma_p} \tag{4}$$

where γ and γ_p are the apparent density of the foamed plastic and the density of the starting unexpanded (monolithic) polymer.

In general, the true (γ^*) and apparent (γ) densities of foamed polymers are different, since γ^* stands for the polymer mass per unit of volume whereas γ denotes the summed polymer and gas masses in the same unit volume. Evidently, in general, $\gamma^* < \gamma$, and only in vacuum $\gamma^* = \gamma$.

Normally, foamed plastics are three-phase systems since they always contain some moisture adsorbed from the environment. We may write:

$$\vartheta_p + \vartheta_w + \vartheta_g = 1 \tag{5}$$

where ϑ_p, ϑ_w, ϑ_g are the volume fractions of the polymer, water and gas, respectively, in the foamed plastic; morlover

$$\vartheta_p = \frac{m}{V_0\gamma_p} = \frac{\gamma}{\gamma_p}; \quad \vartheta_w = \frac{m_w - m}{V_0\gamma_w}; \quad \vartheta_g = \frac{m_g}{V_0\gamma_g} \tag{6}$$

where V_0 is the volume of the foamed plastic; γ_p, γ_w, γ_g, are the densities of the polymer matrix, water and gas, respectively; γ denotes the volumetric weight of the foamed plastic; m, m_w, m_g are the masses of the dry sample, the wet sample and the gas in the foamed plastic, respectively.

In practical calculations ϑ_g is defined as

$$\vartheta_g = 1 - \vartheta_p - \vartheta_w \tag{7}$$

In general $\vartheta_g < G$, since parameter G represents the relative geometrical volume of the voids whereas ϑ_g is the gas-filled portion thereof, since the remainder is occupied by moisture, i.e. $\vartheta_g + \vartheta_w = G$, and only for dry samples $\vartheta_g = G$.

3.3 Rule of Reciprocals

We now introduce the filling ratio U which represents the ratio of the volume of the solid (polymer) phase V_s to the total volume of the system V_0, i.e. to the sum of volume of the gas-filled voids V_g and that of the substance V_s:

$$U = \frac{V_s}{V_g + V_s} = \frac{V_s}{V_0} \tag{8}$$

From this definition it follows:

$$G + U = 1 \tag{9}$$

$$0 \leq G \leq 1 \tag{10}$$

$$0 \leq U \leq 1 \tag{11}$$

From relations (10) and (11) it follows that G and U are always positive and dimensionless, their numerical values lying between zero and unity. The first limiting relation $(G = 0)$ corresponds to a solid, monolithic body, where $U = 1$: The second limiting relation $(G = 1)$ corresponds to a "maximum" porous body, a slid, mono-second limiting relation $(G = 1)$ cornesponds to "maximum" porous body, whose elements are so small that we may take $U = 0$. The relation (9) indicates that G is always complementary to unity in the formula for U. These properties suggest mathematical probabilities. Therefore, formally, we may regard G and U as parameters standing for certain probabilities [4].

What is actually the meaning of the parameters G and U with respect to probability? The value U is the probability of the filling of a finite volume of an addition system by the elements of a porous system or, alternatively, the probability of encountering individual elements within a given volume. Accordingly, G is the probability of the recurrence of voids in a system. It is extremely important that, as might be expected for dimensionless probability parameters, neither G nor U depend on the nature of the elements or the absolute geometrical dimensions of the latter [2].

In an analysis of the morphology, the introduction of the parameter G makes sense only for classification purposes, since the numerical values of G are quite unrelated with the structural characteristics of a system. Apparently, each structure, each type of packing has a corresponding specific value of the gas-filling parameter G. However, the reverse is generally not true, i.e. each particular value of the gas-filling (porosity) parameter may correspond to a large number of different packings, even for elements of the same shape, and of course an infinite number of packings of elements of different shapes.

From the definition (9) it follows that theoretically both parameters G and U are equivalent, as might be expected for the probabilities of two contrary events (substance — void). Since the values G and U are mutually complementary to unity, it is possible to formulate for porous systems a "rule of reciprocals" [2]. If there is a porous system where the volume of voids is V_g and the volume of solid matter V_s, an "exchange" of these parameters (V_g "becoming" V_s) would yield a new porous system in which the former G would assume a new value U. In this second theoretical system the voids would correspond to the solid matter of the former, i.e. G ⇄ U. The "reversal" of gas-filled structurs may conveniently be expressed by the relation

$$\zeta = \frac{G}{U} = \frac{G}{1-G} = \frac{1-U}{U} \tag{12}$$

Eq. (12) has commonly been used, e.g. in the analysis of mass and gas transfer in gas-filled systems. From this relationship it may be deduced that ζ varies from 0 to ∞ and that $\zeta = 1$ when G = 0.5. Systems for which G $\ll$ 0.5 are "poorly gas-filled" ("low-porous") and those with G $\gg$ 0.5 are "highly gas-filled". The rule of reciprocals ("reversal rule") facilitates the analysis of gas-filled structures such as foamed plastics by enabling the use of the so-called complementary gas-filled (porous, cellular) systems. The complementary systems relate to each other as a "mold and casting" or "negative and positive".

The chief advantage offered by this rule, proposed by Radushkevich [2], is not only that it considerably reduces the number of systems being subject to statistical treatments but also simplifies the latter. For instance, by applying the "reversal rule" to closed-cell foamed plastics we convert the latter into their reciprocals in which the individual particles (formerly cells) are arranged in such a manner that they have very few contacts, if any. The latter systems formally all exhibit the characteristics of suspensions (or sols) which can at present readily be studied and described using the well-developed statistical apparatus, which is not possible for the original real system (foamed plastic), due to the insurmountable mathematical difficulties involved.

3.4 Problems

In the physics of solids, especially cellular and porous materials, there are two problems which should be studied separately: 1) the quantitative description of the body and 2) the elucidation of the mechanisms of the physical and physico-chemical processes occurring in the body. Up to now, these two problems have not been clearly differentiated. Thus, physical processes such as the mechanical loading of foamed plastics could not be described correctly since it has so far only been possible to describe a porous macrostructure qualitatively. On the other hand, a macrostructure cannot be depicted qualitatively on the basis of the processes involved. It is extremely difficult to break through this vicious circle since experimental investigations of macrostructures also involve a number of special methodological difficulties. Actually, the data on the physical processes taking place in such systems reveal that these processes are strongly affected by the morphology and are perhaps unique to porous polymers.

The statistics of porous systems belong to the statistical geometry [5], since statistical methods are used here for the treatment of three-dimensional geometrical structures. Mathematically, this approach is similar to the concept of ideally ordered sphere-packing geometries [6]. At present, several approaches to a statistical treatment of porous media are being developed: empirical methods [9,10], molecular analogies, elementary statistical methods [11-17], and general statistical-physical methods [18-21].

4 Fundamentals of Closest Sphere Packing Theory

4.1 Packing Types

In order to obtain a clearer understanding of the morphology of foamed plastics
a review of some basic concepts of classical crystallography, especially the closest
sphere packing theory, may be useful. In fact, the geometrical problem of the
maximum filling of a space by spheres helps to gain a clearer insight into the various
cellular structural types of foamed plastics, provided that in the case studied the
solid spheres are substituted by hollow spheres (an idealized model of foamed plastic
GSE), and the free spaces between the spheres by the continuous polymeric medium.
Such substitution is possible by the reversal rule formulated above for porous structures.

Spatially symmetrical packing is usually based on sphere packings on a plane. There are two
symmetry types, cubic and hexagonal. In a planar arrangement there is only one way of close
packing (Fig. 1, a). In order to obtain a closest packing in the second layer, each sphere of this
layer could be interposed between any three spheres of the first layer. In both packing types the first
two layers have the same arrangement. The packing pattern is different only in the third layer.
Three-dimensional closest packed structures are generated by stacking closely packed layers so
that the spheres are located over voids of the previous layer. Since there are two voids for each sphere,
there are two possibilities of placing the third layer. In the cubic closest packing (Fig. 1, b), the third
layer is placed such that it covers the group of voids which is not covered by the second layer
(symbol ABC). In the hexagonal closest packing (Fig. 1, c), the third layer is placed in the same way as
the first layer (symbol ABA).

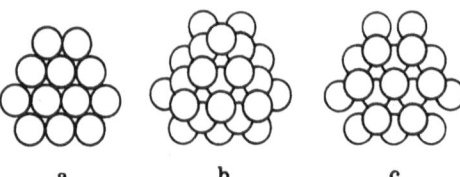

a   b   c

Fig. 1a—c. Planar arrangements of sphere
packings; (a) closest packing in the first
layer; (b) and (c) resulting cubic and
hexagonal three-dimensional closest packing structures (for clarifying the spheres
are slightly separated)

Of course, the packing density is the same in both cases but the symmetry is
different. At the same time, all types of closely packed spheres have the coordination
number 12 (see Sec. 4.2).

Provided that all spheres are in contact, the spatial requirement of the spheres
will be the same for cubic and hexagonal structures and equal to 74.05 % (see Sec. 3.2).
If the spheres are of different size, the gas-filling ratio may be as high as 85 %. In
reality, the shape of the foamed plastic cells is almost never spherical. Therefore, the
space occupied by cells in such materials may even exceed 95 % (see Sects. 5.2 and 6.1).

The relative volume of the solid phase of idealized cellular structures is independent
of the absolute size of spheres but is determined solely by their spatial type of
packing. In the first approximation, this volume is inversely proportional to the
coordination number, i.e. the number of contacts per cell [17] which for closely
packed structures is 12. The shape assumed by the solid phase in a cellular system
is also defined by the packing type, and for spheres of the same size the solid

particles will exhibit the shape of regular polygons. In reality, however the packing of macroscopic spheres of the same size and shape is less ordered and furthermore inhomogeneous.

4.2 Coordination Number

Tests with metallic spheres revealed that the number of contacts may vary from 4 to 12, the average being 8 to 9 [18-20]. Similar experiments with rigid spheres [21-26] showed that there are two random packing types: a close packing with the filling ratio (see Sect. 3.3) $U = 0.31$ ($G = 0.69$) and a looser packing with $U = 0.41$ ($G = 0.59$). Here, the filling ratio U means the fraction of the volume space between the spheres, and the gas-filling, the ratio G is the volume fraction of the spheres. According to Bernal [5], the "ideal" coordination number of 12 never occurs with rigid spheres, the real value always being lower.

Radushkevich [1] has calculated the radial distribution function W(r) for sphere packing systems by making the following assumptions. Consider two "marked" spheres at a distance x_i. For m pairs of spheres the sum of distances would be $\sum_0^m x_i$. If $m \to \infty$, the probability to find two spheres at the distance x_i is $x_i \Big/ \sum_0^\infty x_i$. Let N be the number of spheres in the system and one of the spheres mark the center of the system, then, the number of spheres at the distance x_i from the central sphere is $(N-1)\, x_i \Big/ \sum_0^\infty x_i$. This calculation permits to compute the number of spheres inside the spherical shell between the distances r_1 and r_2 from the central sphere. In 1 cm³ of such a shell, the number of n is given by

$$n = (N - 1) \sum_{r_1}^{r_2} x_i \Big/ \sum_0^\infty x_i \int_{r_1}^{r_2} 4\pi r^2 \, dr$$

If the packing density in this system is N/V (where V is the overall volume of the system), n/(N/V) is the radial function:

$$W(r) = \frac{V}{N} \frac{3}{4\pi} (N - 1) \sum_{r_1}^{r_2} x_i \Big/ \sum_0^\infty x_i (r_2^3 - r_1^3) \tag{13}$$

Since Eq. (13) contains the packing density, G and U and the coordination number v can be calculatev.

For the filling ratio $U = 0.49$, $v \cong 6$; for $U = 0.6$, $v \cong 4$; for $U = 0.74$, $v \cong 1–2$. Hence, it follows that for $U > 0.7$ the spherical system becomes virtually gas-like and $W(r) = 1$.

Irrespective of the different procedures used, the reported results are in fair agreement. In particular, it is always possible to represent, though roughly, a gas-filled structure as a superposition of a random and a closest (cubic) sphere packing system. Besides, sphere packing data may prove very useful in the investigation of the morphology of many gas-filled plastics which, according to the "rule of reciprocals" (cf. Sect. 3.3), are the counterparts of packed sphere systems. For foamed plastics, the radial distribution function may also be expected to be similaar to the one found, for example, for liquids [12,23] since the dense portions have imaginary centers with the contact numbers 11 to 12 which are surrounded by looser portions with contact numbers of 4 to 6 [24-26].

However, the real morphology of foamed polymers may only roughly be simulated by the packing types considered in the preceding section. Real foamed plastics are chiefly characterized by a structure composed of a broad range of cells with different diameters; furthermore, they contain no cells with regular spherical shape [27-30].

The different foamed plastic structures will be discussed below. Here, is only mentioned that there are at least two descriptions of such structures. The first involves the structure model of the foamed plastic consisting of originally monodispersed spheres. The spaces between the latter are filled with spheres continuously decreasing in size. The second approach involves a calculation of the shape and sizes of the initially monodisperse packing of spheres which change their shape to polygons of different size and shape [17,31-33].

5 Open- and Closed-Cell Foamed Polymers

In order to establish and evaluate quantitatively the relation between the morphology and the properties of foamed polymers the basic macrostructural parameters must be determined. These parameters include: relative number of open and closed cells, volumetric weight or apparent density cell size, shape, wall thickness, cell distribution according to size and shape in a given volume and specific surface area of the foamed plastic material.

We will now estimate the effects of these parameters on foamed polymer properties.

5.1 General Problems

Gas-filled polymers may contain both isolated and communicating gas structural elements. By varying the chemical composition of the polymer phase and the foaming conditions either predominantly closed-cell or predominantly open-cell structures are obtained. From a morphological point of view, according to Blair [34], open-cell structures are only obtained if the following conditions are fulfilled: 1) at least two pores or two broken faces in each spherical or polygonal cell; 2) an overwhelming majority of cell struts must belong to at least three gas structural elements.

These differences in physical structure are responsible for the different properties of foamed polymers containing varying proportions of open cells. In contrast to closed-cell foamed polymers, open-cell foamed plastics have a higher water and moisture absorptive capacity, a higher gas and vapor permeability, less pronounced electric and heat insulation characteristics but a stronger sound absorptive and damping power [35].

In open-cell structures, the gas phase is air whereas in foamed plastics with isolated cells it may in addition contain hydrogen, carbon dioxide, and volatile liquids, depending on the blowing agent employed. Of course, diffusion, processes gradually replace these gases with air, but for some time the original gas phase composition may considerably affect certain properties of the foamed plastic, including heat and electric conductivity, shape stability, etc. In particular, if freons are used as blowing agent, their diffusion rate from the polymer matrix is below that of the atmospheric air to the interior of the foamed plastic, hence, an excess pressure will develop the cells of the material which may ultimately cause the cell walls, to break. Conversely, the higher diffusion rates of hydrogen and carbon dioxide in the polymers compared to that of air may result, for a short period, in a pressure which is below the atmospheric pressure, causing shrinkage or even partial breakdown of the GSE walls.

The use of surfactants can considerably influence the type of the foamed plastic GSE [36-40]. Moreover, the morphology of the obtained foamed plastics is also greatly affected by the blowing technique [41-44].

According to Thomas [45], elastic polymers often produce open-cell foamed plastics, whereas rigid polymers generally form closed-cell materials. However, there are many exceptions to this rule, owing to the variety of blowing techniques. Closed-cell structures are more likely to be produced from polyurethanes, epoxy resins, silicones, poly(vinyl chloride), polystyrene, etc., whereas open-cell materials mainly result from phenolic and carbamide foamed plastics.

5.2 Open Cells and Volumetric Weight

The relation between volumetric weight and the proportion in the polymer of open-cell or closed-cell GSE has not yet been studied. It is only known that for any polymer composition the relative percentage of open GSE increases as the volumetric weight of the foamed plastic decreases. This is due to cell growth involving a decrease of the thickness of cell and struts. It adversely affects the aggregation stability and may ultimately cause fracture of the cell walls.

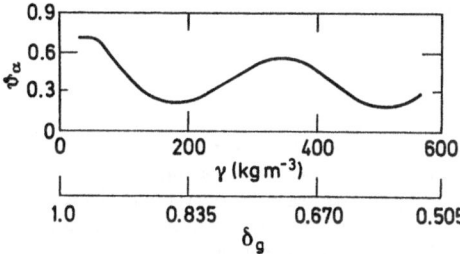

Fig. 2. Relationship between the volume ratio of open gas structural elements ϑ_α and the volumetric weight γ and volume ratio of gas ϑ_g for rigid polyurethane foam (Grade PPU-3)

As the data of Chaikin and Shutov suggest (Fig. 2), the $\vartheta_\alpha = f(\gamma)$ curve always exhibits two maxima, where ϑ_α is the relative content of open GSE and γ the volumetric weight of the cellular polymer. One of the maxima is located in the region of low γ values while the other is found at γ volumes corresponding to the gas-phase volume ratio $\vartheta_g = 0.65 - 0.75$. The gas-filling values G (see Sects. 3.2 and 4.1) corresponding to the closest sphere packings also lie within the same range of ϑ_g values. Evidently, this coincidence is not fortuitous, since in closest packings where the number of contacts between cells is 12, the probability of polydispersed cell fracture according to the above-described mechanism (due to the different internal pressures in gas bubbles of different radii) also increases.

We have pointed out above that the application of foamed plastics is determined by the proportion of open GSE's. However, until very recently, the formulas used for the calculation of this proportion did not take into account the parameter ϑ_α. From experimental data on the moisture absorption kinetics of an expanded rigid polyurethane (PPU-3) with the volume weight $\gamma = 70$ to 700 kg/m³ under normal conditions, Chaikin and Shutov proposed the following relation between the relative water absorption and the parameters of the macrostructure of the foamed plastic, assuming that the cells are spherical:

$$\frac{\vartheta_w}{\vartheta_w^c} = \frac{6\vartheta_\alpha\vartheta_p^2}{2\sqrt{2}(\vartheta_p^c)^3\,\vartheta_\alpha^c}\,\exp\left[-\frac{\vartheta_p^2}{2(\vartheta_p^c)^2}\right] \tag{14}$$

where ϑ_p, ϑ_p^c are the respective volume ratios of the polymer in a given sample and in a closely
  packed sample;
  ϑ_α, ϑ_α^c are the volume ratios of open cells in a given sample and in a closely packed
  sample;
  ϑ_w, ϑ_w^c are the volume ratios of water in a given sample and in a closely packed sample.

We should now consider the factor "6" in the numerator of the pre-exponential in Eq. (14) which has the physical meaning of either a form factor, packing factor or coordination number of the closest spherical (cubic) sphere packing. The points of Fig. 3 are somewhat scattered probably because of the equal proportions of open cells (ϑ_α) in samples of different volume weights. Eq. (14) has a maximum when $\vartheta_p/\vartheta_p^c = 1$ (Fig. 3), i.e. when the volume ratio of polymer in a sample is equal to that in a plastic foam with closest spherical packing; the gas phase volume is then 74% which, for polyurethane foam, corresponds to a volumetric weight of 315 kg/m³.

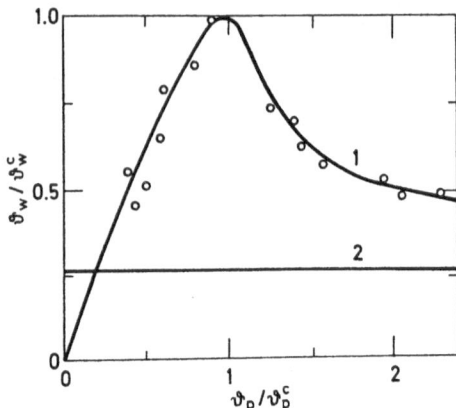

Fig. 3. Relationship between moisture sorption and structural parameters of foamed polymers, see Eq. (14); (1) rigid polyurethane foam grade PPU-3; (2) unfoamed polyurethane, the same grade

5.3 Reticulated Foams

There are various secondary processing methods permitting the finished closed-cell foams to be converted to open-cell foams by breaking the cell walls. This is achieved by hydrolysis, oxidation, application of elevated or reduced pressure, heat or mechanical treatment [41-43]. By such a modification an original macrostructure may be converted, for example, into a reticulated plastic foam, in which the cells are totally devoid of walls and the polymer phase is contained exclusively in the GSE struts (Fig. 4). Such materials have a very low density ranging from 3 to 10 kg/m³ and a relative solid phase volume of 0.3 to 1.0%.

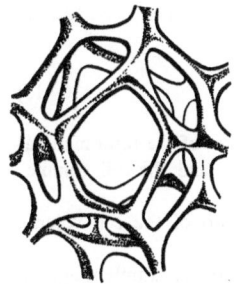

Fig. 4. Schematic view of the gas structural element of rigid resol phenolic foam (sketch of microphotograph, magnification 150 times)

All available methods of preparing reticulated plastic foams are divided into chemical and physical methods [43]. In the chemical methods, a finished plastic foam is subjected either to alkaline or acid hydrolysis which destroys the walls more rapidly than the thicker struts. This method has been commercially utilized for the production of reticulated polyurethane foams from polyesters.

In the physical methods the cell walls are destroyed by physical forces, usually by mechanical load. According to one of the most widespread and cheapest methods [46], the walls are broken through blast-wave impact. To this effect, the blowing gas is removed from the foamed plastic and the material is saturated with a gaseous explosive (e.g. a stoichiometric O_2/H_2 mixture, an air/H_2 mixture, acetylene, or atomic oxygen) in a special chamber where blast occurs under a controlled pressure (up to 2 atm). Other physical methods suitable for preparing reticulated foams use a thermal shock or repeated compression cycles.

The resultant reticulated foams are used as air filters in engine exhaust systems, fibrous heat insulation and damping materials, etc.

6 Volumetric Weight of Foamed Polymers

6.1 Classification

The volumetric weight, a parameter characteristic for the relative contents of both the solid and gas phase in a material, is the fundamental morphological parameter of foamed polymers and is related to all relevant physical properties of foamed plastics such as strength, thermophysical and electric properties.

Modern cellular polymer technology allows foamed plastics with a very wide range of volumetric weights γ to be prepared from 3 to 900 kg/m³, i.e. with gas-filling values G varying from namely 99.5 to 10% (see Sects. 3.2 and 4.1). Depending on their γ value, all cellular plastics may be divided into five classes: very light ($\gamma = 3 - 50$ kg to m³), light ($\gamma = 50 - 200$ kg/m³), medium ($\gamma = 200 - 500$ kg/m³), heavy ($\gamma = 500 - 700$ kg/m³), and superheavy ($\gamma > 700$ kg/m³) weights. In a first approximation, it may be assumed that most of the physical characteristics of polymer foams are directly related with the volumetric weight [37,47]. This very general, assumption holds true both for elastic [37,48-50] and rigid foams [46,51].

In general, the volumetric weight is determined by the true densities of the polymer phase and the gas phase. It is related with the gas-filling of the foam as follows [36]:

$$\gamma = \gamma_p(1 - G) + \gamma_g \tag{15}$$

where γ_p is the true density of the polymer phase which is equal to the ratio of the total material mass to the difference between the total volume of the material and the cell volume;
γ_g is the gas density in the cells and
G is the gas-filling (porosity) of the foam, being equal to the ratio of the cell volume to the total material volume.

In practive, the volumetric weight of plastic toams is calculated as the ratio of sample weight Q to its geometrical volume V_0:

$$\gamma = \frac{Q}{V_0} \tag{16}$$

where the volumetric weight is given in kg/m^3 or g/cm^3.

6.2 Volumetric Weight and Cell Size

The relation between the volumetric weight and the mean cell diameter is described by a hyperbolic function. This function has been analytically derived by Romanenkov et al. [36] from the following geometrical model.

If we imagine a foamed plastic structure consisting entirely of spherical cells of diameter d arranged in a cubic lattice, then for a wall thickness of δ the weight of the polymer (g_c) per cell will be:

$$g_c = \gamma_p \left[\frac{4\pi}{3} \left(\frac{d + \delta}{2} \right)^3 - \frac{4\pi}{3} \left(\frac{d - \delta}{2} \right)^3 \right] = \frac{\pi}{3} (3d^2\delta + \delta^3) \, \gamma_p \tag{17}$$

where γ. is the density of the polymer phase.

The number of cells (N_b) in a cubic space with a side "b" is $N_b = (b/d)^3$. The polymer weight within this space can then be expressed by

$$g = N_b g_c = \frac{4}{3} (3d^2\delta + \delta^3) \, (b/d)^3 \, \gamma_p \tag{18}$$

Since the volume $V = b^3$, the volumetric weight of the foamed plastic is:

$$\gamma = \frac{g}{V} = \frac{\pi}{3} (3d^2\delta + \delta^3) \left(\frac{b}{d} \right)^3 \frac{\gamma_p}{b^3} = \frac{\pi(3d^2\delta + \delta^3)}{3d^3} \gamma_p \tag{19}$$

and, disregarding δ^3, we finally have:

$$\gamma = \frac{\pi\delta\gamma_p}{d} \tag{20}$$

If we assume that $\delta = $ const, Eq. (20) will present a gcod approximation of the relation between the volumetric weight and the mean cell diameter within small volumetric weight ranges. A plot of relation (20) at constant wall thickness $\delta = 1$ micron for a PSB polystyrene foam yields the curve shown in Fig. 5. The deviation from the experimental result was in this case 22–28%.

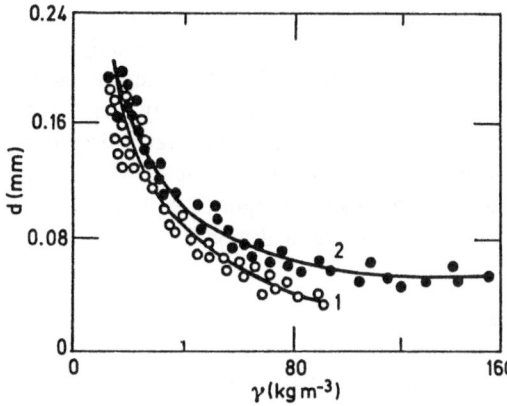

Fig. 5. Relationship between the volumetric weight of polystyrene foam, γ, and mean diameter d of macrocells; (1) and (2) curves plotted according Eq. (20) for grades PSB and PSB-S respectively [36]

6.3 Volumetric Weight and Wall Thickness

Within a sufficiently broad range of volumetric weights the wall thickness is not constant but a linear function of the volumetric weight:

$$\delta = \delta_{cr}(1 + \alpha\gamma) \tag{21}$$

where δ_{cr} is the critical (minimum) wall thickness which is still capable of supporting a cellular structure;

α is a parameter specific to the given polymer.

Substituting Eq. (21) into Eq. (20) and equating $d_{cr} = \pi\alpha\delta\gamma_p$ we get:

$$\gamma = \frac{\pi\delta_{cr}\gamma_p}{d - d_{cr}} \tag{22}$$

For a stable foam structure, the basic cell parameters such as diameter and wall thickness have to be within a certain range. Thus, according to Romanenkov, the limiting conditions of the applicability of Eq. (22) are given by the following inequation:

$$d_{cr} \leqq d \leqq d_{max} \tag{23}$$

where d_{cr}, d_{max} are the minimum and maximum cell diameters, respectively, of a stable liquid polymer foam.

The experimentally determined cell diameters depend not only on the type of the starting polymer but also on the composition of the blowing agent and the foaming process. Thus, the critical parameters of the cellular structures of unpressed polystyrene foams of grades PSB and PSB-S are:

Parameter:	d_{cr}, μm	d_{max}, μm	δ_{cr}, μm
Grade PSB	26	215	0.85
Grade PSB-S	10	140	0.40

The use of other geometrical models for the analytical treatment of the relation $\gamma = f(\delta, d)$ does not markedly affect the form of Eqs. (20) and (22) but only alters the value of the factor in the numerator of Eq. (22)[36]. For the spherical model discussed above, this factor is $\pi = 3.14$. For other models it may vary from 3.0 to 3.3. In contrast to the spherical model, other models involve greater discrepancies between the experimental and theoretical results. For example, when using a cubic model for PSB plastic foam having a volumetric weight of 90 ± 0.5 kg/m³, the discrepancy is 18 % as compared with 9 % in a spherical model.

6.4 Volumetric Weight Distribution

The "volumetric weight" is an averaged value because within the volume of a plastic foam block, especially across its depth, the volumetric weight of the surface layers may be 3 to 10 times greater than the averged value γ. The non-uniformity of the density distribution in plastic foams is chiefly due to the processing conditions. Thus in expanded molten polymer stock, non-uniformity in temperature distribution may result in a different distribution of the decomposition products of the blowing agent, across the different portions of the molten material[52, 53]. On the other hand, the decrease of γ from the bottom to the top of a foamed product, which invariably occurs in practice in all polymers and for any foaming technique, is an inherent feature of the foaming process, especially with the decrease in thickness of the GSE wall and the strut, due to the draining of the still liquid foam. There is, however, one exception to this rule: γ is uniform in articles prepared from pre-expanded granules of e.g. polystyrene foam. In this technology, draining occurs prior to rather than during the formation of the foamed article, particularly during the pre-expansion of the granules. The two steps, pre-expansion and final expansion, may be separated by dozens of hours. Indeed, the bar chart and γ density distribution in PSB polystyrene foam (Fig. 6) show that the volumetric weight is distributed uniformly, and the Pirson test verifies that its distribution is normal. When the volumetric weight distribution is other than normal, the asymmetry and excess coefficients of the corresponding curves may vary between 0.1 to -0.4 and 0.02 to 0.6, respectively, for different kinds of plastic foams[44].

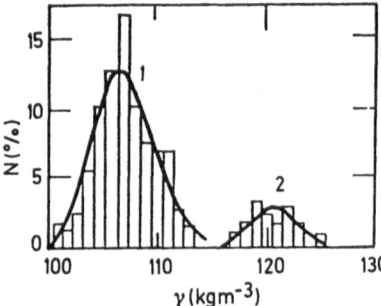
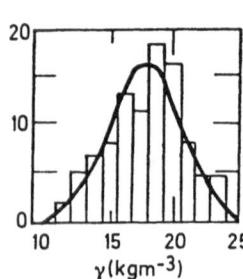

Fig. 6a and b. Volumetric weight distribution of (**a**) polyurethane foam and (**b**) polystyrene; (1) intermediate (core) layer; (2) surface layer [36]

The mold material has a marked effect on the density distribution pattern in a foam block. It has been shown [54] for the grade FRP-1 of phenolic foam in panels with aluminum or asbestos cement claddings that, due to the different thermoconductivity coefficients of the claddings, the thickness of the compacted layer may be 1.5 to 2.0 cm for aluminium and 0.5 to 1.0 for asbestos cement. For phenolic foams the variation of the volumetric weight across the depth of the intermediate layer of a three-layer panel may be described by the following equation (Fig. 7):

$$\gamma_y = \gamma_{min} + \gamma_1 \{\exp(-\Delta\gamma y) + \exp[-\Delta\gamma(l - y)]\} \tag{24}$$

where γ_y is volumetric weight of the foam at a distance "y" from the surface
     γ_{min} the minimum the minimum volumetric weight
     γ_1 the volumetric weight increment near the interface of the cladding intermediate layer
     $\Delta\gamma$ a "compaction gradient" characterizing the rate of volumetric weight variation across the direction of compaction and
     l   the intermediate layer depth (expansion space).

γ_{min} and γ_1 are obtained from the relations $\gamma_{min} = 0{,}44\bar{\gamma}$ and $\gamma_1 = 16{,}3\bar{\gamma}^{0.6}$, where $\bar{\gamma}$ is the averaged volume weight of the foam block. The compaction gradient is 2.0 for asbestos cement and 0.70 for aluminum cladding layers. The calculated and experimental $\Delta\gamma$ distribution curves are in good agreement (Fig. 7).

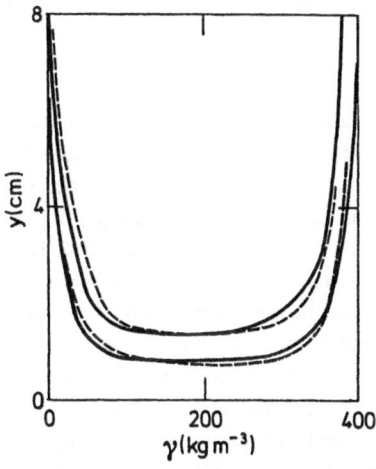

Fig. 7. Volumetric weight distribution over the depth h of resol phenolic foam inside the "sandwich" panel: (1) $\gamma = 70$, (2) $\gamma = 116$ kg/m³; ——— experiment, ——— calculation [54]

6.5 Problems

The actual volumetric weight of plastic foams depends largely on the amount of the blowing agent used. However, for each type of blowing agent there is a maximum concentration. If this concentration was exceeded, the volumetric weight of the foam would not be reduced further, but the polymer strength would be deteriorating, due to plastification of the polymer by the degradation products of the blowing agent and due to the unusually rapid pressure rise in the GSE.

In general, the volumetric weight of foamed plastics is a complex function of chemical (type of polymer and blowing agent) and physical (foaming technique and conditions applied) factors. The same volumetric weight characteristics may, for a given polymer, be attained either by controlling the blowing agent concentration or by controlling the physical conditions of the foaming process (for a given blowing agent concentration). The first approach is commonly preferred in practice, i.e. the desired volumetric weight is adjusted by controlled amounts of the blowing agent.

Up to now, the second approach has not been utilized effectively by technologists. Of course, this approach is relatively difficult to realize in practive, since it would require of the process engineer to abandon the purely empirical approach to the problem and undertake a profound investigation of the physico-chemical expansion process parameters. However, the second approach would allow to save blowing agent which is often rather expensive and not readily available.

7 Cell Shape of Foamed Polymers

7.1 Cell Shape Models

The macrostructure of a real dispersed material, including plastic foams, cannot be visualized without recurring to a geometrical model; hence, the concepts of the material structure are always conventional, being valid only on the basis of the GSE model used. Nevertheless, such concepts are necessary both for a quantitative description of a macrostructure and for a comparative analysis of the morphological parameters of the different cellular plastics.

In applications to a real foam the term "shape" as used here is also quite conventional. When speaking of a GSE shape some statistically averaged characteristic is usually meant. This characteristic is determined from measurements of a large number of cells by statistically treating the measured results in an appropriate manner, or by visually estimating the shape of the majority of foam cells. In the first case, the cell shape is characterized quantitatively (see below) while in the second qualitatively (sphere, ellipse, polygon, etc.). A real foam structure, however, even within a very limited volume of material, contains cells with a large number of shapes and dimensions. The true picture is so complicated that so far no methods for calculating the GSE shape distribution have been proposed while there are several methods for calculating the cell size distribution function (see Chapter 9). That is why all qualitative treatments of the relationship between the application properties of plastic foams and the GSE shape (or size) are based on a "statistically averaged" cell shape, or on idealized and always simplified models of real cells. In mathematical analyses of the plastic foam morphology the real structures have been simulated by the following idealized shapes: monodispersed spheres [52-55], spheroids [56], cubes [54,57,60], hexagons [58,61], rhombic dodecahedrons [62], oblong pentagonal dodecahedrons [63,64], complex polyhedrons [57], capillaries [69] and other [59,65].

7.2 12- and 14-Hedrons

From the liquid foam stability theory it follows that a liquid foam is most stable when the gas bubbles are strictly spherical in shape since, according to the Laplace and Plateau laws, the interface area and the capillary pressure have minimum values in this case. As it has been shown in Sect. 4.1 for the monodispersed

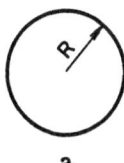 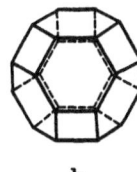 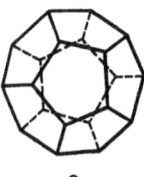

a   b   c

Fig. 8a—c. "Ideal" cell geometries; (a) sphere; (b) 14-hedron; (c) 12-hedron. A 14-hedron is composed of 6 squares and 8 hexagons, the surface of a 12-hedron is composed of regular pentagons only [66]

spherical cellular structure, the closest packing pattern is attained at a gas phase volume of 74%, each sphere contacting 12 neighbors. On further increase of the gas phase volume ratio the spheres tend to become polygons, in the ideal case dodeca-hedrons with pentagonal faces.

However, the geometrical parameters of a regular 14-hedron are more susceptible to the formation of an "ideal" foam structure than 12-hedrons (Fig. 8), since they are closer to the geometry of a regular sphere. Thus, for a given volume, the surface areas of sphere, 14-hedron and 12-hedron are scaled as 1:1.06:1.10. The fact that real foam structures more frequently assume a dodecahedron shape is due to their somewhat larger surface area which is a regular pentagon; in contrast, the inequality of angles in a 14-hedron promotes unbalanced capillary pressure and therefore produces a foam structure which is more susceptible to coalescence.

According to Harding's scheme [66] a combination of three contacting bubbles of equal size (two in front and one in the rear) can give rise to the following morphological types of cell walls and struts (Fig. 9). The first type is spherical packing of the bubbles which may be realized in high-density foamed plastics where the gas phase occupies less than 74% of the total volume. The second type, a structure consisting of absolutely regular dodecahedrons, is impracticable since the capillary pressure in the struts of such a structure would be infinitely high (due to the zero-curvature radius). The third structural type may be realized in closed-cell light-weight foams where the gas phase occupies more than 74% of the volume. Draining tends to deform the dodecahedron faces whereby the curvature radius of the struts becomes non-zero. The extent to which the real shape deviates from that of a regular dodecahedron depends on polymer viscosity and surface tension and, for the same polymer, on the expansion process conditions. As the gas phase volume increases, the cell walls become thinner and the liquid phase drains to the struts under the effect of the resultant capillary pressure drop. A certain portion of the walls is broken, resulting in foam coalescence.

The fourth type, an "open" dodecahedron, may be observed open-cell foams, provided the viscosity of the starting polymer phase is high enough. If one sixth of all walls, or more, have been broken, the resultant plastic foam will be an entirely open-cell foam, i.e. feature the so-called "through"-porosity. The fifth morphological type (called "web" structure) is a minimum surface dodecahedron having all the material of the cell walls drained off to the struts.

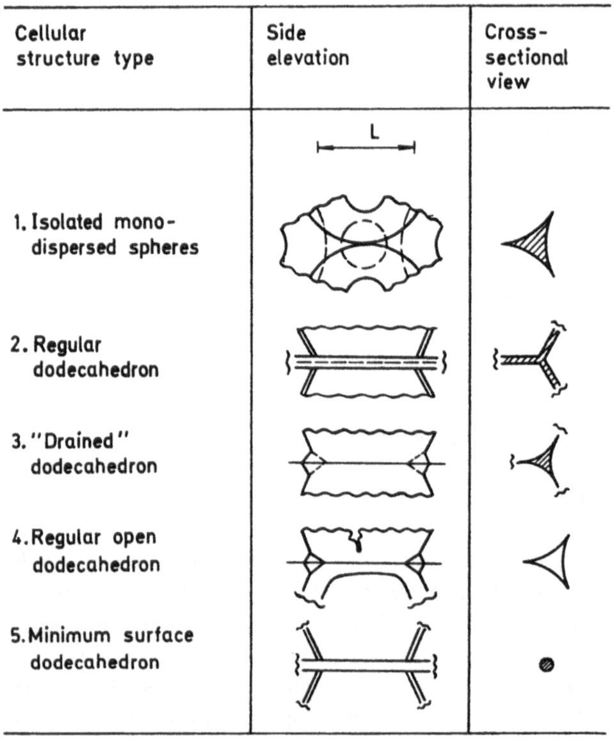

Cellular structure type	Side elevation	Cross-sectional view
	L	
1. Isolated mono-dispersed spheres		
2. Regular dodecahedron		
3. "Drained" dodecahedron		
4. Regular open dodecahedron		
5. Minimum surface dodecahedron		

Fig. 9. Possible types of structures for walls and struts of gas structural elements with the same length L, according to Harding [66]

The last type is identical with the foam structures obtained by means of the reticulation methods (see Sect. 5.3), the difference being only due to the mode of preparation.

The available expansion theories still inadequately predict all the varieties of the particular morphological structures occurring in real foams. Nevertheless, Harding's model based on the dodecahedral structure concept covers a broad range of the really occurring structures. In particular, as the microscopic investigations have shown, the cell shapes of most closed-cell foamed plastics are quite close to that of a "drained" dodecahedron, the degree of draining (the amount of polymer draining from the walls to the struts of the cells) being higher for oligomeric than for high polymeric foams and smaller for rigid than for flexible foams. The cell and strut shapes of open-cell foamed polymers closely resemble the structure of an open dodecahedron (Fig. 9). Cracks and cavities occur more often in the cell walls of rigid than in flexible plastic foams.

Most cell walls are pentagon-shaped, the remainder being squares and hexagons which are approximately equal in number. Giuffria [67,68] suggests that the cells of a rigid polyurethane foam may contain as many as 15 faces, the shapes of which range from triangular to octagonal.

7.3 Anisotropy of Cell Shapes

7.3.1 General Problems

The major discrepancy of the morphology of most real foamed plastics from that predicted by theory is the cell orientation in the direction of foaming, resulting in an oblong form of the cells. This is due to the fact that the mechanical stresses generated during foaming are not distributed uniformly throughout the entire volume of the plastic foam; thereby the gas bubbles tend to expand along the minimum local stress directions [70]. Moreover, there are usually several such directions within a given volume of the plastic foam, running at different angles with respect to the base plane of the casting. Furthermore, for the same direction the aspect ratio (ratio of largest to smallest linear dimension) of cells may be widely different. Therefore, when carrying out morphology investigations, the material sections have to be studied not only at different block depth (or width) positions but also in the different planes relative to the basal plane.

Stretching of the substantially dodecahedral cells is schematically illustrated in Fig. 10 (a regular dodecahedron is projected on the two planes perpendicular to Fig. 8 as a hexagon and on that of Fig. 8 as a decagon).

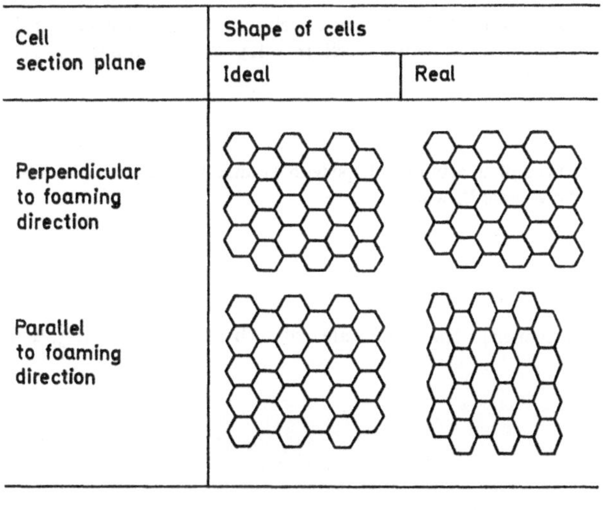

Cell section plane	Shape of cells	
	Ideal	Real
Perpendicular to foaming direction		
Parallel to foaming direction		

Fig. 10. Schematic view of cell elongation during the foaming process [66]

Stretching of cells along the foaming direction is always more pronounced in plastic foams produced by free-rise foaming (as opposed to expansion in a closed space), and is the stronger, the lower the volumetric weight of the material. In free-rise foaming, the elongation of cells decreases from the bottom to the top layers of the block, the upper layers consisting, as a rule, substantially of spherical cells. In the expansion in a closed space the cellular anisotropy of plastic foams is always lower than in free-rise foaming, and cell elongation occurs only near the mold walls. The higher the pressure developed in the mold, the closer to spherical the shape of the major portion of cells. However, the outer layers of a foam article usually contain very small cells, and under high pressure the surface layer of a plastic foam block is normally smooth [68,69].

7.3.2 Roundness and Sphericity of Cells

When analyzing the cell shape, one must differentiate between the two different notions roundness and sphericity. In a general case, the roundness φ is defined as

$$\varphi = \frac{\sum_i r_i^c/R_i^c}{B} \tag{25}$$

where r_i^c is the curvature radius of the planar projection of the ith-cell
$\qquad R_i^c$ maximum radius of the inscribed circle in this cell,
$\qquad B$ the number of angles in the polygonal cell face.
When the corners are rounded off, r^c tends towards R^c and in the limit at $r^c \quad R^c$ the roundness reaches a maximum ($\varphi = 1$) only ideal cirumference. The different types of roundness are shown in Fig. 11 [19].

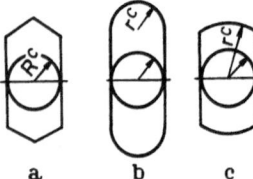

Fig. 11a—c. Different types of cell roundness: **a** polygonal cell, **b** maximum roundness cell ($r = R$), **c** small roundness cell ($r > R$) [71]

To define sphericity let us imagine a cell of an arbitrary shape with wolume V_c and surface area S_c. The diameter of a spherical cell of the volume V_c will be

$$D_n = \left(\frac{6V_c}{\pi}\right)^{1/3} \tag{26}$$

where D_n is called the nominal cell diameter. The surface area of such a cell is $S_s = \pi D_n^2$, and the ratio

$$\psi = \frac{S_c}{S_s} \tag{27}$$

is the actual definition of the cell sphericity. It must be understood that $\psi = 1$ only for strictly spherical cells, otherwise $\Psi < 1$ [69,71].

7.3.3 Cell Elongation

Cell elongation in the direction of expansion, which is responsible for the anisotropy of the properties of plastic foams may be quantitatively evaluated using the cell aspect ratio, i.e. the ratio of the largest to the smallest dimension of the cell. Like any other quantitative shape or size characteristic of polydisperse systems such as foamed polymers, these ratios are averaged values. The smaller the size, the more difficult it becomes to evaluate the cell aspect ratio [19,64].

Cell (or particle) aspect ratios are conventionally divided into volume ratios (α_V) and surface ratios (α_S). If the measured cell diameter is (for instance, D is the maximum linear cell dimension along or across the foaming direction), then the colume V and surface area S of such cell will be given by [71]

$$V = \alpha_V \frac{\pi}{6} D^3 ; \quad S = \alpha_S \pi D^2 \qquad (28)$$

where α_V is volume aspect ratio and
$\quad \alpha_S$ surface aspect ratio.

The numerical expressions of α_V and α_S depend on the adopted simplified cell model and, with the exception of the simplest cases, cannot be calculated a priori. For spherical cells the ratio $\alpha_S/\alpha = 6$ [72,73].

The cell aspect ratio, which is characteristic for foam anisotropy, can be determined by microscopically measuring the linear cell dimensions on sections cut from different portions of a foam block. The results of such measurements can only be representative of the morphology of certain portions of a cellular structure. A statistical analysis of samples cut of different block portions and in different directions with respect to that of expansion is a very difficult task.

7.3.4 Influence of Cell Shape on Mechanical Properties

The cellular anisotropy of foamed plastics must inevitably affect their strength, thermophysical, dielectric and other properties. For example, the mechanical strength characteristics of foams which display structural anisotropy are markedly different along and across, always being better along the foaming direction. Thus, if the cell height-to-width ratio (h/D) is increased from 3.5 to 5.3, that is by a factor of 2.8, the compressive strength of polyurethane foams also increases 2.8 times, the tensile strength increasing 3 times and shear, bending strengths and Young's modulus 2 times. For brittle foams, these values are slightly lower. An increase of the test temperature, suppresses the effect of the h/D ratio on strength characteristics (Fig. 12) [69].

Very interestingly, Harding [66] found that the compressive strengths of polyurethane and polystyrene foams in one case and of polyurethane and phenolic foams in another case display the same dependence on the height-to-width ratio over a range from 1/2 to 2/1, despite the fact that the volumetric weights of these materials are

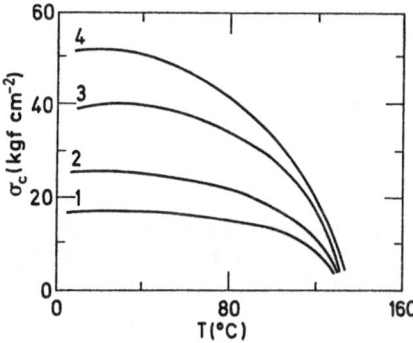

Fig. 12. Effect of temperature on compressive strength σ_c at different height-to-width ratios h/D of polyurethane foam cells; aspect ratios: (1) 3/5, (2) 1/1, (3) 5/3 [67]

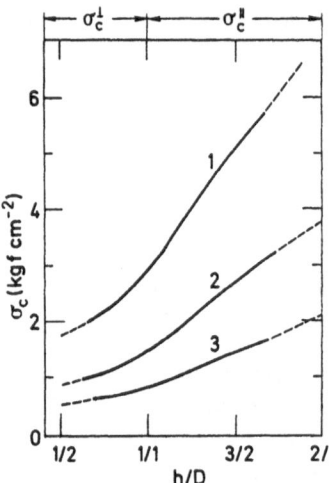

Fig. 13. Effect of height-to-width ratio h/D of cells on compressive strength of rigid foamed polymers based on: (1) polyurethane, $\gamma = 48$ kg/m^3, polystyrene, $\gamma = 32$ kg/m^3; (2) polyurethane, $\gamma = 32$ kg/m^3; (3) polyurethane, $\gamma = 24$ kg/m^3; phenolic resin, $\gamma = 32$ kg/m^3; $\sigma_c^\perp$ and $\sigma_c^\parallel$ are the compressive strengths perpendicular and parallel to the direction of foaming [66]

different (Fig. 13). From these results it may be concluded that in certain cases the anisotropy of foam morphology may have an even more marked effect on the strength characteristics of plastic foams than chemical nature of the polymers, volumetric weight or degree of cell closeness. Probably, the equal strengths of cells so different both chemically and physically is due to differences in the polymer material distribution between the cell walls and struts. Since the compressive strength of polymer foams is substantially dependent on the GSE strut stiffness, the GSE struts may be thinner and the walls thicker in polystyrene and polyurethane foams than in phenolic foams.

In rigid polyethylene foam ($\gamma = 32$ kg/m^3, cell diameter between 0.5 and 1.5 mm) the anisotropy of the macrostructure is particularly reflected by the shape of the stress-strain diagram (Fig. 14). When a load is applied normally to the foaming direction, the deformation of the material increases perceptibly. In contrast, the resistance of the structure to compressive stress applied in the direction parallel to foaming increases since unit surface area of the material contains more rigid GSE struts in the latter case than in the former.

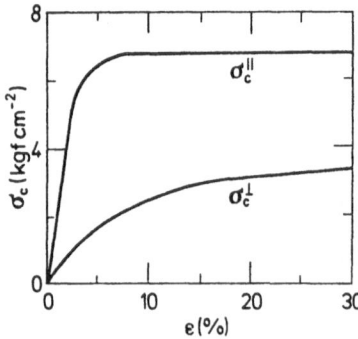

Fig. 14. Effect of load, either parallel $\sigma_c^\parallel$ or perpendicular $\sigma_c^\perp$ to the direction of foaming, on the compressive strength of polyethylene foam $\gamma = 32$ kg/m^3 [97]

7.4 Ultrasonic Method for the Estimation of Cellular Anisotropy

7.4.1 Basic Relationships

A promising method of controlling and evaluating the structural parameters of plastic foams is the acoustic (ultrasonic) pulse method [74, 75]. This method is based on the propagation of elastic waves in a porous medium depending on shape, distribution and volumetric content of the GSE.

Theoretical premises for the determination of the physicomechanical properties by the acoustic pulse method were derived from the following three models of real foam structures [74, 75]:

1) an isotropic structure in which the cells are nearly spherical;

2) a structure of a transversally an isotropic medium in which the cells have the shape of an ellipsoid of revolution and are oriented parallel to the foaming direction, i.e. perpendicular to the porous medium surface;

3) the structure of an orthotropic medium in which the cells are oriented normal to the foaming direction, i.e. parallel to the surface, and have the shape of a three-axial ellipsoid with different dimensions along the symmetry axes.

The cellular anisotropy of plastic foams may be evaluated by e.g. an anisotropy coefficient q which is equal to the ratio between average cell dimensions along the major symmetry axes of the respective model. An isotropic material 1s characterized by only one anisotropy coefficient ($q = 1$), a transversally anisotropic material by two ($q_1 = q_2$ and q_3) and an orthotropic medium by three coefficients ($q_1 \neq q_2 \neq q_3$). The effective anisotropy coefficient q^* which characterizes the anisotropy of the whole material is given by:

$$q^* = \frac{2v_3}{v_1 + v_2} \tag{29}$$

where v_1, v_2 and v_3 are elastic wave propagation velocities along the cell orientation axes in a plastic foam.

According to Karapetyan [74], the coefficient q^* is, in turn, related with the elastic wave propagation parameters in a cellular material and foam morphology by the following formula:

$$\bar{v} = q^* p \bar{v}_p \vartheta_p + \bar{v}_g \vartheta_g \tag{30}$$

where $\bar{v}$, $\bar{v}_p$, $\bar{v}_g$ are elastic wave propagation velocities in plastic foam, polymer matrix and air, respectively, ($v_g = 330$ m/s)

ϑ_g and ϑ_p are the polymer and gas phase volumetric contents in plastic foam; for simplicity, it is assumed that $\vartheta_g + \vartheta_p = 1$, cf. Eq. (5);

p is a parameter accounting for the effect of cell shape and size on the elastic wave propagation parameters.

As G. and S. Olivé [75a] correctly indicated, the definition of the effective anisotropy coefficient q^* in Eq. (29) is unfortunate because the case $q^* = 1$ may be given not only for isotropic foamed polymers but also for anisotropic ones (for example, if $v_1 = v_3 - x$ and $v_2 = v_3 + x$; see Table 1 for an example).

A more reliable definition of the anisotropy is by means of experimental determination of some properties along the major symmetry axes. For the acoustic pulse method it means measurement of the wave propagtion velocities v_1, v_2 and v_3 along the cell orientation asces. In this case, the real anisotropy coefficients q_1, q_2 and q_3 are calculated as (see Table 1)

$$q_i = \frac{v_i - v_g \vartheta_g}{pv_p(1 - \vartheta_g)} \tag{30a}$$

where $i = 1, 2, 3$ correspond to the symmetry axes.

7.4.2 Anisotropy of Foamed Polymers

As shown by Karapetyan in [74, 75], the structural anisotropy coefficients differ very widely not only between different types of plastic foams but even for one and the same plastic foam grade, depending on its volumetric weight. Table 1 shows that the velocity of ultrasonic waves as well as the mechanical strength characteristics are always higher in the direction of expansion. The anisotropy is most pronounced in press-molded polystyrene foam of grade PS-4 in which the cells are elongated in the direction of the pressing force [36]. In contrast, unpressed polystyrene foam of grade PSB-S which is manufactured by thermal shock has an isotropic structure. The data of Table 1 for polyurethane foam of grade PPU-3 support the well-known law, invariably observed in mechanical tests, that the foam anisotropy tends to decrease with increasing volumetric weight. Thus, the anisotropy of plastic foams seems to be related with cell orientation and their distribution in the material.

8 Cell Size of Foamed Polymers

8.1 Methods of Estimation

The characterization of the plastic foam cell size by a linear dimension (often referred to as cell diameter) is simple, convenient and now commonly applied. However, a single linear dimension can unambiguously describe only geometrically regular cells, i.e. spheres and ellipsoids. In all the other cases studied it remains to be defined what is meant by linear dimension.

As with cell shapes of a real foam, cell sizes in this material can also be characterized only by nominal (effective) values. The actual effective values depend, first, on the observation method (whether direct — macroscopic, or indirect — adsorption, volumetric, picnometric, etc.). Secondly, they depend on the particular simplified model of the structure and cell shape and thirdly on the method of processing the measured data.

If direct measurements are used and followed by statistical processing of the measured results, the cell size is conventionally represented via the so-called mean cell diameter which is defined differently by various authors [71]. If there are N cells of the same shape but of different diameters D_i, the mean probable cell diameter will be a number $\bar{D}_p$ which has the following property: one half of all cells have diameters below $\bar{D}$ and the other half above it. Among other mean cell diameter definitions the more generally used are: arithmetical mean diameter $\bar{D}_a = \Sigma D_{p_i}/N$, geometrical mean diameter $\bar{D}_g = \sqrt[N]{D_{p_1}, D_{p_2} \cdots D_{p_N}}$, mean surface diameter $\bar{D}_s^2 = \Sigma D_{p_i}^2/N$; mean volume diameter $\bar{D}_v^3 = \Sigma D_{p_i}^3/N$ and others (for details see Sect. 10.2).

Table 1. Anisotropy of Plastic Foam Structures From Acoustic Measurement Data [74, 75]

Foam type	Volume weight γ (kg/m³)	Gas-filling factor G (%)	Parameter P	Longitudinal Wave Velocity			Anisotropy coefficients (Eq. 30a)			Effective anisotropy coefficient q* (Eq. 29)
				v_1	v_2	v_3	q_1	q_2	q_3	
Polystyrene,										
Grade PSB-S	52	95	5.0	900	890	890	1.0	0.98	1.0	1.0
Grade PS-4	40	96	6.0	910	890	1350	1.0	0.97	1.09	1.51
Phenolic, Grade FRP-1	62	95	5.0	920	870	890	1.0	0.93	0.97	1.0
	77	94	3.5	860	910	860	1.0	1.11	1.0	0.97
	105	91	3.5	1070	1050	900	1.0	1.0	0.75	0.75
Polyurethane, Grade PPU-3	57	95	4.5	870	850	710	1.0	0.96	0.71	0.83
	70	94	4.0	840	780	680	1.0	0.84	0.66	0.83
	170	86	2.2	1000	990	880	1.0	1.0	0.83	0.88
	200	83	2.0	960	1060	970	0.88	1.0	0.90	0.96
	255	79	1.6	1070	1080	980	1.0	1.02	0.90	0.91
	372	69	1.4	1300	1290	1310	1.0	0.99	1.0	1.01
	467	61	1.4	1340	1350	1330	1.0	1.0	1.02	1.01
	510	58	1.2	1380	1390	1420	1.0	1.0	1.03	1.02

Alternatively, the cell size may be characterized by the hydrodynamic radius which is equal to the ratio of the cross-sectional area of the cell to the perimeter of crossection [76], or by a "nominal" cell diameter equal to that of the largest sphere circumscribing the cell [70]. The cell size is characterized in an indirect manner in terms of specific surface, thermal conductivity coefficients, air and moisture permeability, etc.

However, a real foam structure is composed of cells having differing shapes, sizes and volumes. In studying the properties of foamed polymers as well as in developing and elaborating preparative processes, it is necessary to find out cell size, shape and volume distribution. The methods for calculating the respective distribution functions will be discussed in Sect. 9.2, 9.3; here, we only note that the cell size distribution function is a most comprehensive and valuable characteristic of plastic foam structures.

8.2 Cell Size and Number of Cells

Normally, 1 cm³ of an isotropic plastic foam contains 1,000 to 10,000 cells, their sizes being statistically distributed in the bulk of the material so that the deviation of the real size from the mean cell size is within the range from 12 to 25 % (rarely up to 50 %) [66, 69]. Cell sizes and cell size distribution depend not only on the polymer grade but also on the foaming process conditions (for more information see Sect. 9). The number of cells per unit volume of foam (N_c) is an important and frequently used parameter, especially in estimating the effectiveness of a nucleating agent or the macrostructural homogeneity throughout the sample bulk. The parameter N_c is a function of cell size and volumetric weight of the plastic foam and, according to Fehn [79], is given by:

$$N_c = (1 - \gamma/\gamma_p) \bigg/ \frac{\bar{D}^3}{6000} \tag{36}$$

where N_c is the number of cells per cm³ of foam,
    γ the volumetric weight of foam (g/cm³),
    γ_p the polymer matrix density (g/cm³), and
    $\bar{D}$ the mean cell size (mm).

An interesting investigation of plastic foam structures has been proposed by applying the modern concepts of the incipience, growth and structure of inorganic polycrystals to the problems of morphology, formation mechanisms and stability of foamed polymers [81, 82]. Comparing the typical elementary cells of a plastic foam and a polycrystalline metal, the faces and struts of the former do not, like the latter, have portions with rapidly varying curvature radii (Fig. 15) and the volume distribution of cells is also greatly different. Thus for a plastic foam the number of cells N_V having a volume V is given by the expression

$$N_V = N_{V_{max}} \exp\left[- k(V/V_{max} - 1)^2\right] \tag{37}$$

where V_{max} is the most frequently observed cell volume,
    $N_{V_{max}}$ the number of cells having this volume, and
    k   a numerical parameter.

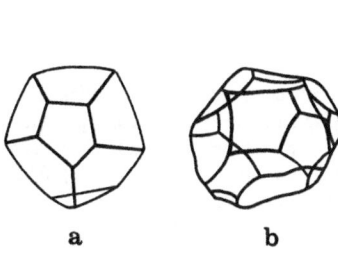

0.8

$\overline{D}$ (mm)

0.4

0

3   5   7   9

n

Fig. 15a and b. Schematic views (a) cell of rigid polyester foam and (b) magnesia polycrystal [82]

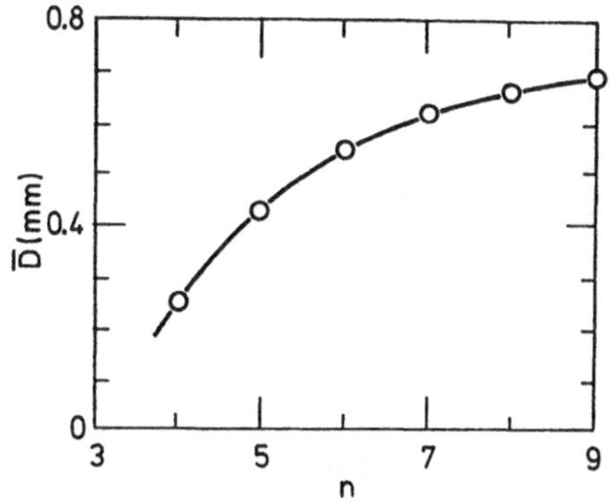

Fig. 16. Mean cell diameter $\overline{D}$ as a function of the number of planes n in the polygons of polyester foam cell sections [81]

The foam cell diameter depends on the number of faces (Fig. 16) and, on the average, is proportional to their volume cubed. The cell size distribution in a cut-off section is never symmetrical, and the maximum value of this distribution will be a diameter larger than the mean cell diameter. It must be noted that the above distribution law depends on the assumed range of cell sizes which manifests itself in the numerical value of the factor k in Eq. (37).

8.3 Cell Size and Wall Thickness

The volumetric weight and the ratio between the number of open and closed cells are the fundamental morphological parameters of foamed plastics. Nevertheless, it has been reported that even for the same volumetric weight and number of open (or closed) cells the strength and thermophysical parameters may be widely different in plastic foams made of the same polymer grade. The differences in cell shapes and sizes are responsible for this fact.

A structure in which the same volumetric weight can be achieved with different cell sizes is attainable by the following relation. It is known that for monodisperse spherical cells of a plastic foam the general relation between mean wall thickness δ and mean cell diameter $\overline{D}$ is given by Aleksandrov's formula [80]:

$$\delta = \overline{D} \left(\frac{1}{\sqrt{1 - \gamma/\gamma_p}} - 1 \right) \tag{38}$$

where γ_p is the monolithic polymer density (kg/m³) and
γ the volumetric weight of plastic foam (kg/m³).

If $\gamma_p = 1000$ kg/m³, than the wall thickness in a foam of volumetric weight $\gamma = 1000$ kg/m³ at $\overline{D} = 0.2$ mm will be 8 microns. A plot of δ versus $\overline{D}$ yields a straight line (Fig. 17). From this relation it follows that for a given polymer the same wall thickness or the same cell size may be obtained at different gas phase contents. Thus, the volumetric weight of a plastic foam may be controlled not only

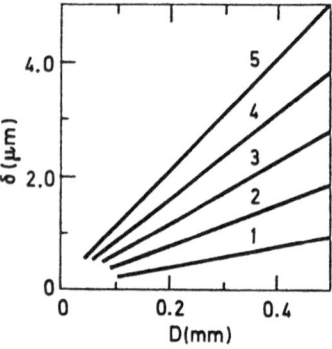

Fig. 17. Relationship between average cell wall thickness δ and mean diameter of spherical cells D̄ of foamed polymers at different ratios of the volumetric weight of foams γ and density of unfoamed polymer γ_p; γ/γ_p: (1) 0,05; (2) 0,10; (3) 0,15; (4) 0,20; (5) 0,25 [80]

by varying the cell size but also by varying the thickness of its cell walls. Conversely, foams of the same volumetric weight may have cells of different size.

8.4 Microcells

8.4.1 Types and Sizes of Microcells

We have already mentioned in Sect. 6.2, 6.3 that the physicochemical conditions of the foaming process and the foam stability criteria determine the upper and lower limits of cell sizes so that, depending on the polymer type, composition, and foaming process conditions, the upper limit of size may be as large as a few millimeters [36,83-85]. Until recently, it was believed that the minimum size of a plastic foam cell cannot be less than several dozens of microns (Table 2). However, by the application of scanning electron microscopy and the mercury penetration method, plastic foam structures were found to incorporate gas voids whose minimum dimensions were fractions of a micron, i.e. 2 or 3 orders of magnitude smaller than could be observed earlier in cellular polymers [7,8].

Table 2. Foam Macrostructure Parameters [36]

Foam Type	Polymer density γ_p (kg/m³)	Volumetric weight γ (kg/m³)	Mean cell diameter (mm)
Polystyrene	1050	160–220	0.02–0.2
Poly(vinyl chloride)	1380	50–220	0.1–0.3
Polyurethane	1200	50–500	0.1–0.5
Phenol-formaldehyde	1200	20–200	0.2–0.5
Urea-formaldehyde	1450	8–20	0.2–0.5

Morphological structures as isolated or communicating microspheres and their conglomerates (or "windows") in macrocellular walls were observed by scanning electron microscopy by Shutov et al. [86-88] in phenolic foams. The reviewer termed them "microcells" [7,8]. These results were corroborated by Lowe, Barnett, Chandley and Dyke [89]. Later, microcells were found in oligomeric foams of other kinds as well (polyurethane foams [88] and multicomponent plastic foams based on urethane-phenolic oligomers [92]). Thus, in an investigation of polyurethane grade PPU-3 foams by optical microscopy of thin sections (with transmitted light) two groups of cells could be observed, macrocells of a few dozen to a few hundred microns in size and microcells of 0.01 to 1 micron in size; the latter were about 2 to 3 orders of magnitude more numerous than macrocells (Fig. 18). Investigations of similar samples by mercury penetration tests verified these results and revealed that the cell size distribution function is not Gaussian (standard) but logarithmically-normal (Fig. 19). Such a distribution pattern indicates [71] that the foam cells consists of several, at least two groups, the cells of each group being radically different as regards size and number.

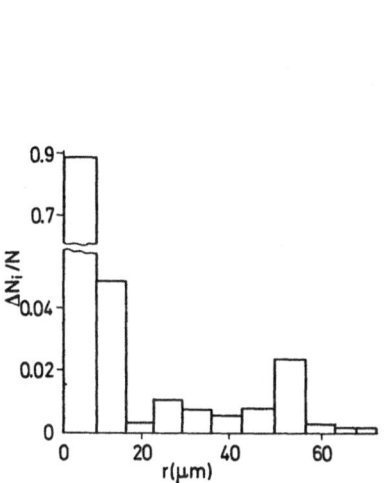

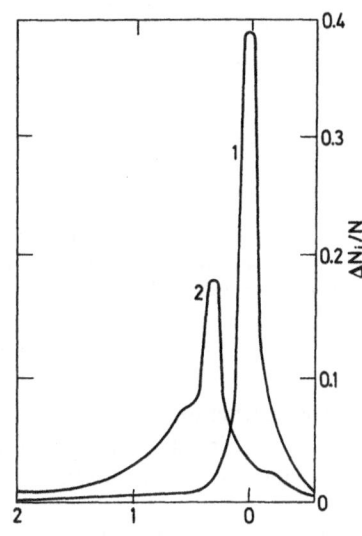

Fig. 18. Bar chart of cell size distribution in rigid polyurethane foam $\gamma = 130$ kg/m³ (grade PPU-3) according to optical microscopy data [89, 94]

Fig. 19. Cell size distribution of rigid polyurethane foam (grade PPU-3) samples having the volumetric weight (1) 40 and (2) 500 kg/m³ according to mercury penetration data [94]

From the data (Fig. 20) of the author it follows that microcells are the most frequently occurring form of gas voids in foamed plastics based on reactive oligomers. Another characteristic feature of microcells is that their absolute size is practically independent of the volumetric weight of the foam; when the latter is varied from 40 to 500 kg/m³, the mean size decreases only by a factor of 2 (from 1 to 0.5 microns). A third feature of microcells is their shape: the bulk of microcells are spherical whereas most macrocells are oblong (see Chapter 7.3).

On the assumption that the packing of microcells is closest and their shape is spherical, it is possible to estimate the wall thickness in such structures using

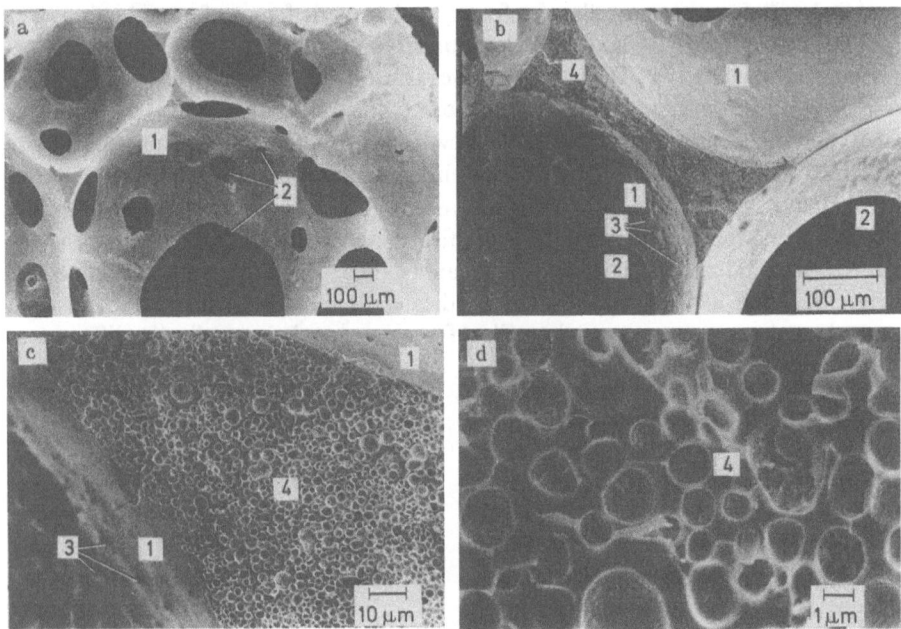

Fig. 20a—d. Cellular structure of rigid phenolic foam (grade FL-1) according to scanning electron microscopy data: **a—d** magnification 48, 200, 1000 and 6000 times respectively; (1) macrocells, (2) macroholes in the macrocells walls, (3) microholes in the macrocells walls, (4) microcells in the strut of macrocells

Eq. (38). The calculation yields $\delta = 0.1$ micron in fair agreement with electron microscopic data (Fig. 20). The wall thickness of microcells is only one order of magnitude smaller than their radius.

8.4.2 Matrix Model of Cellular Structure

These results reveal that a plastic foam structure may be considered as a system of thin films and, therefore, support a model of plastic foam morphology namely a matrix system composed of thin polymeric films defining two groups of cells (macro- and microcells). Additional support in favor of this model of plastic foam structure is provided by the studies on the electric properties of plastic foams [93, 94]. Among the numerous equations so far advanced for the calculation of the dielectric properties, the expressions which describe the dry foam structure by one of the limiting cases of a matrix system, namely a laminated dielectric structure with layers parallel to the force lines of the electrical field, agree best with the experiments [8, 88].

To conclude the discussion we want to underline that the microcells in a foam structure should not be regarded as a defect of its morphology due to a faulty preparation or improper formulation. Rather, an analysis of the foaming process and the morphology of foamed plastics based on reactive oligomers demonstrates unambiguously that microcells belong to the foaming process due to the disturbance of the dynamic balance of the oligomer-gas system during the final stages of the formation of the cellular structure from self-expanding casting compositions. This problem is analyzed in detail in a special publication [88] and in a review [8].

It should finally be noted that such microstructures have not been detected in thermoplastic foams. However, in such materials the cell size distribution also fails to follow some standard pattern (with the exception of a special case of foams prepared from pre-expanded granules), but is rather logarithmically-normal; this fact indirectly indicates of the existence, also in these materials, of at least two groups of cells of widely different sizes [91].

8.5 Cell Size and Physical Properties

Cell size is a factor controlling not only the volumetric weight of a foam but also such an important morphological parameter as gas-filling (see Sect. 3.2). As follows from Fig. 21, when the cell size decreases (i.e. the number N of cells per milli- meter increases, the gas-filling of polyurethane foam samples (2.5 cm thick) also decreases since the number of the closed cells increases. This has been confirmed by air stream resistance tests at a pressure drop of 0.25 mm H_2O [41].

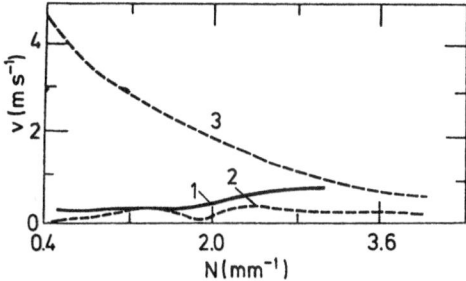

Fig. 21. Gas permeability of foamed polymers as a function of cell size, in the coordinates: air steam velocity v versus number of cells N per linear mm, in polyurethane foams based on polyethers (1) and polyesters (2, 3): (1), (2) closed cell foams, (3) reticulated foam [41]

All other conditions being the same (chemical composition, volumetric weight, closeness of cells), the cell size can considerably affect the properties of foamed plastics. For instance, the thermal conductivity coefficient of foams always increases when the cell size increases due to the growing number of radiation and convection paths of heat transfer [38, 95, 96] (for more information see Sect. 12). An increase of the cell size also causes a rise of Young's modulus of both flexible [38, 41, 45] and rigid plastic foams [97, 98].

The relation between the cell size and the strength of phenol-formaldehyde plastic foams shows that, for a mean cell diameter of less than 0.2–0.3 mm, their compressive strengths and Youngs moduli increase considerably as compared to foams with large cells. It has been found that for many flexible foam types the tensile strength and ultimate elongation are the higher the smaller the cells [99, 100].

The effects of open- and closed-cell structures of foams on their strength indices was studied by Blair[41] for standard, non-reticulated ($\gamma = 32$ kg/m^3) and reticulated polyurethane foams (the latter by hydrolyzing the cell walls with caustic soda). The experiment was aimed to study simultaneously the effect of the presence or the absence of walls in cells of various size on the strength characteristics. When the cell size is increased, the compressive strength of plastic foams decreases, this is evident under high compressive loads (Fig. 22, curves 1 and 2). The compression of flexible non-reticulated plastic foams involves bending of the struts and stretching of the walls (Fig. 23). On further compression, the concomitant tensile stresses tend to break the cell wall forming in it a kind of "window". Tearing the cell wall does not, however, spread over to the cell struts. In a reticulated structure, the compressive stresses only bend the cell struts; therefore, the materials of this type are subject to greater deformation than non-reticulated materials.

Upon stetching flexible foams behave totally different (Fig. 23): reticulated structures are stronger than non-reticulated ones. This fact is all the more unexpected

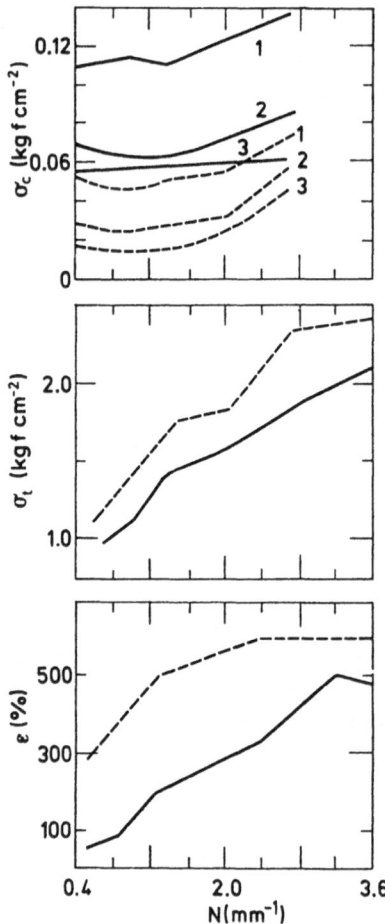

Fig. 22. Effect of cell size (in terms of the number of cells N per linear mm) on the compressive σ_c and tensile σ_t strength and relative elongation ε for flexible polyurethane foams; (1), (2), and (3) denote deformation of compression (%) 66, 50 and 25 respectively; solid lines refer to nonreticulated foams, dashed lines to the same foams with reticulated cells[41]

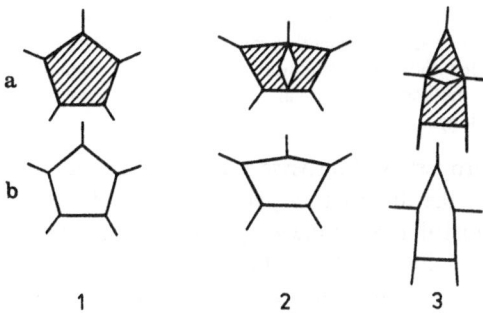

Fig. 23a and b. Cell shape variation (a) in nonreticulated and (b) in reticulated flexible polyurethane foams under compressive and tensile loads: (1) initial state, (2) compression, (3) tension [41]

since reticulated plastic foams contain about 5 to 10% less of the polymeric material (the hydrolyzed walls) than nonreticulated. Blair attributes this fact to the specific patterns of the tensile stress distribution in the structtures. When a non-reticulated material is subjected to tension, the cell wall material also experiences tensile stresses while that of the struts experiences bending stresses. Stretching of a flexible cell results in elongation whereby the cross-sectional area of the wall decreases and the wall material may yield already under relatively small tensile stresses. The fractures and cracks that develop in the walls propagate into the struts and promote their destruction. The reason for the higher strength of the reticulated structures is therefore that in this case only stretching and bending of the cell struts takes place (Fig. 23).

9 Cell Size Distribution of Foamed Polymers

9.1 General Problems

The cell size distribution pattern depends essentially on the foaming techniques and not on the composition of the ingredients (polymer type and content and type of blowing agent). Many types of the so-called integral (structural) plastic foams, i.e. foams with hardened skin layer, can be obtained in a single forming operation from a formulation also, suitable for the preparation of "standard" plastic foams, i.e. foams featuring uniform cell size and volumetric weight distributions throughout the material. To achieve this it appears sufficient to modify only the foaming process parameters, either, in the simplest case, through, injection of molten polymer into a cooled mold (in injection molding processes) or to ensure rapid cooling of the extruded material. On the other hand, for a given polymer the size of its cells will depend both on the blowing agent concentration and the foaming techniques applied.

As already noted in Section 8, it is practically impossible to measure the shape and size of each cell individually therefore, on uses statistical geometry methods for the calculation of the morphological parameters. Originally, these methods had been developed and used in crystallography, petrography and colloid chemistry for analyzing the macrostructures of crystals, metal alloys, minerals, suspensions, etc. [107-109].

For instance, Johnson and Fullman [110,111] reported a graphical analytical method for the evaluation of the true size of the circles observed on a cutaway surface of a porous body.

Reid [112] developed analytical tools for studing the cell size distribution from dissected sphere size data measured on a section surface. The first to introduce a general theory of cell size distribution in a solid body was Gellings [113] who calculated the macrostructural parameters of a number of real metal systems. On the basis of Gellings' method, Mihira et al. [29] developed the principles of the statistical analysis of plastic foam morphology.

In investigations of the structure and properties of disperse materials, particularly plastic foams, it is necessary to find out the distribution pattern of one phase in the material. It is simpler to consider the distribution of the gas phase in a solid body, i.e. the gas-filled cell distribution within a foam bulk] In their turn, the cells, as shown above, may be characterized by several parameters (size, shape, volume and surface area). The cell size distribution pattern is the most comprehensive characteristic of the dispersity of the gas structural elements of plastics. Furthermore, the cell size can, be determined by one of the methods which will be discussed below, and the foam dispersity is expressed in terms of the nominal cell diameter distribution function.

9.2 Fundamentals of Calculation

Statistically, the cell size distribution has been studied for many types of foamed plastics. In most cases, the plots of such functions have a sharp asymmetrical maximum (steep towards smaller cells and smoother towards larger cells, see (Fig. 19).

The asymmetrical cell size distribution functions may be described by the following equation [114]:

$$\frac{dN}{dD} = F\left[\frac{(D - D_{min})(D_{max} - D)}{D_{max} - D_{min}}\right] = F(D) \tag{39}$$

where D   is a particle size (normally the nominal diameter) varying in the range $D_{min} < D < D_{max}$,
         F(D) a logarithmically-normal function.
    The solution of Eq. (39), which is referred to as a non-ideal logarithmically-normal law, is relatively difficult. If we set $D_{min} = 0$ and $D_{max} = \infty$, the cell size distribution will become logarithmically-normal:

$$F(D) = AD^{-1} \frac{1}{\sqrt{2\pi}\sigma} \exp\left[-\frac{1}{2}\left(\frac{\ln D/\eta}{\sigma}\right)^2\right] \tag{40}$$

where A is a normalization constant, η a distribution median, and σ a distribution scattering.
    On the same assumptions regarding the limits of cell sizes, the differential distribution function may be approximated by the following exponential formula

$$F(D) = AD^m \exp(-\alpha D^p) \tag{41}$$

where m, α, p are parameters accounting for the maximum sharpness and the curve asymmetry.
    To find the normalization constant A, the above formulas must be normalized under the following condition:

$$\int_0^\infty F(D) \, dD = 1 \tag{42}$$

For equation (41) the normalization yields:

$$A = p\alpha^{\frac{m+1}{p}} \left[\Gamma \left(\frac{m+1}{p} \right) \right]^{-1} \tag{43}$$

where $\Gamma \left(\dfrac{m+1}{p} \right)$ is a gamma-function.

When an optical or electron microscope is used, the cell diameter is determined indirectly by measuring the dissected cell dimensions. When other data on the cell shape are not available, the determination of the true cell size from section measurements is a very complex, if at all solvable, task.

For a near-spherical cell shape Lenz [115] proposed a calculation which allows the true radii of real cells (r) to be determined from the mean value and the standard deviation of section radius (S).

Let $F(r) = \varrho(r)dr$ be the number of cells in a foam whose radii are within a range from r to (r + dr), and $F(s) = g(s) ds$ be an average number of circular sections on a cut surface whose radii are within a range from s to (s + ds). Then, the following general relations are valid:

$$g(s) = 2s \int_s^\infty \frac{\varrho(r)\, dr}{s\sqrt{r^2 - s^2}} \tag{44}$$

$$\varrho(r) = \frac{1}{\pi r} \int_r^\infty \frac{sg(s)\, ds}{\sqrt{s^2 - r^2}} \tag{45}$$

To find the integrals (44) and (45) we must preset the form of the distribution functions g(s) and $\varrho(r)$. According to Lenz [115], the most general form of function $\varrho(r)$ is

$$\varrho_n(r) = Cr^n \exp\left[-(n+1)r/r_m \right] \tag{46}$$

where C is a constant and
        n an integral number.

For large n this distribution becomes normal (Gaussian), and for r = 0, $\varrho_n(r) = 0$. Depending on the preset value of n, formula (41) will give different distribution functions $\varrho(r)$ with a mean value of r_m, and the function g(s) will then look like

$$g(s) = 2s^{n+1}C(-1)^n \frac{d^n K_0(x)}{dx^n} \tag{47}$$

where $K_0(x) = 0{,}5\pi i H_0^{(1)}(ix)$
is the zero-order Hunkel function and $x = (n+1)s/n_m$

These expressions show that even in the simplest case of spherical cells the mathematical relationship between the functions $\varrho(r)$ and g(s) is complex enough. The discrepancy between them is particularly large for small section radius s. This is all the more unfavorable since cells of large size may produce sections of small diameters [116-118].

9.3 Mihira's Method

To calculate the foam cell size distribution function we consider, following Mihira [29], an isolated cell foam structure model (Fig. 24). Let r be a true cell radius, and s the radius of sectional circles on the cut surface X, $\bar{r}$ and $\bar{s}$ their mean values, σ_r^2 and σ_s^2 their mean square deviations, and f(r) and f(s) their distribution functions. We will denote by x the depth of a cell dissected by the plane X (Fig. 24) and calculate the probability P(r,x) of cells having a radius in the range from r to (r + dr) and a depth from x to (x + dx). The probability P(r) for the cells dissected by the plane X to have a radius r is:

$$P(r) = r f(r) \, dr / \bar{r} \tag{48}$$

and the probability of the cells dissected by the plane X having a depth x is:

$$P(x) = dx / 2r \tag{49}$$

Then the probability P(r, x) will be the product of probabilities P(r) and P(x):

$$P(r, x) = P(r) \, P(x) = f(r, x) \, dx \, dr = f(r) \, dr \, dx / 2\bar{r} \tag{50}$$

provided that:

$$0 < x < 2r; \qquad 0 < r < \infty \tag{51}$$

From Fig. 24 it follows that $s^2 = x(2r - x)$; the upper limit for s is $\sqrt{x(2r - x)} < a$ and, therefore, the probability P(s) at s $\leq$ a is:

$$P(s \leq a) = \int_{x(2r-x)<a^2} f(r, x) \, dr \, dx \tag{52}$$

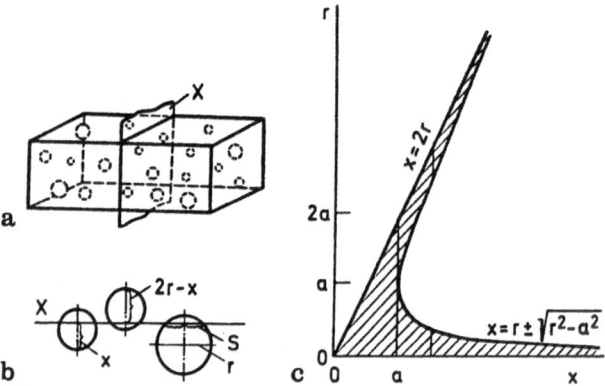

Fig. 24a—c. Model of foamed polymer with isolated cells: (a) block dissected by plane X; (b) relationship between cell radius r and radius of circles s on section surface; (c) calculation of functions f(r) and f(s) under limiting conditions: $0 < x < 2x$; $x^2 - 2rx + a^2 = 0$ [29]

The integration limits of Eq. (52) correspond to the hatched area in Fig. 24c:

$$P(s \leq a) = \frac{1}{2\bar{r}} \int\limits_{x(2r-x) < a^2} f(r) \, dr \, dx = \frac{1}{2\bar{r}} \left[\int\limits_0^a \left(\int\limits_0^{2r} f(r) \, dx \right) dr + \right.$$

$$\left. + \int\limits_a^\infty \left(\int\limits_0^{r-\sqrt{r^2-a^2}} f(r) \, dx + \int\limits_{r+\sqrt{r^2-a^2}}^{2r} f(r) \, dx \right) dr \right] =$$

$$= \frac{1}{\bar{r}} \left[\int\limits_0^\infty rf(r) \, dr - \int\limits_a^\infty \sqrt{r^2 - a^2} \, f(r) \, dr \right] = 1 - \frac{1}{\bar{r}} \int\limits_a^\infty \sqrt{r^2 - a^2} \, f(r) \, dr \qquad (53)$$

For a number of theoretical distribution functions of sphere radii f(r), the authors [29] calculated the distribution of the radii of circles on the sectioned surface, making use of Eq. (53). The result is shown in Fig. 25, (for comparison, the maximum cell diameter, b, was taken as equal in all calculations).

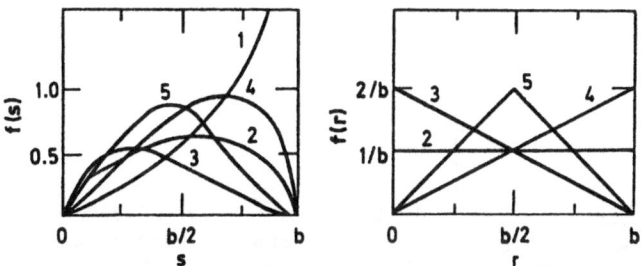

Fig. 25. Relationship between functions f(s) and f(r) for different cell distribution patterns in foamed polymer: (1) uniform size cells; (2) continuous distribution; (3) decreasing distribution; (4) increasing distribution; (5) triangular distribution; here, b is the maximum cell diameter [29]

Mihira et al. also calculated the statistical mean, $\bar{x}$, as well as variance ϱ_x^2 and coefficient of variation $\varkappa$, according to

$$\bar{x} = \int\limits_0^b xf(x) \, dx; \qquad \sigma_x^2 = \int\limits_0^b x^2 f(x) \, dx - (\bar{x})^2; \qquad \varkappa = \sigma_x / \bar{x} \qquad (54)$$

with $x = r$ or s respectively. The results are given in Table 3. It turned out that $\bar{s}$ can either be smaller (e.g. for equal spheres $\bar{s} = 0.785 \, \bar{r}$) or larger than $\bar{r}$ (e.g. for decreasing distribution $\bar{s} = 1.178 \, \bar{r}$). It should be noted that the value of the average radius of the cells may be in error by 10–20 % if the difference between f(r) and f(s) is not taken into account. Similar errors can arise for variance and the coefficient of variation.

In order to illustrate the different types of the function f(s) that can be obtained from the same cellular polymer but at different packing factors (see Sect. 4), the authors [29] prepared some cellular poly(vinyl alcohol)polymers foamed with nitrogen. The

Table 3. Statistical morphological parameters of r and s for cellular polymers [29]

Parameters	Foam cell distribution pattern				
	Equal sphere	Continuous	Decreasing	Increasing	Triangular
$\bar{r}$	b	b/2	b/3	2b/3	b/2
σ_r^2	0	$b^2/12$	$b^2/18$	$b^2/18$	$b^2/24$
$\varkappa_r$	0	$1/\sqrt{3}$	$1/\sqrt{2}$	$1/2\sqrt{2}$	$1/\sqrt{6}$
s	$\pi b/4$ $= 0.784\,\bar{r}$	$\pi b/6$ $= 1.047\,\bar{r}$	$\pi b/8$ $= 1.178\,\bar{r}$	$3\pi b/16$ $= 0.884\,\bar{r}$	$7\pi b/48$ $= 0.916\,\bar{r}$
σ_s^2	$0.0498\,b^2$	$0.0592\,b^2$ $= 0.7104\,\sigma_r^2$	$0.0458\,b^2$ $= 0.8244\,\sigma_r^2$	$0.0530\,b^2$ $= 0.9540\,\sigma_r^2$	$0.0385\,b^2$ $= 0.9240\,\sigma_r^2$
$\varkappa_s$	0.284	0.465 $= 0.806\,\varkappa_r$	0.315 $= 0.446\,\varkappa_r$	0.391 $= 1.107\,\varkappa_r$	0.336. $= 0.823\,\varkappa_r$

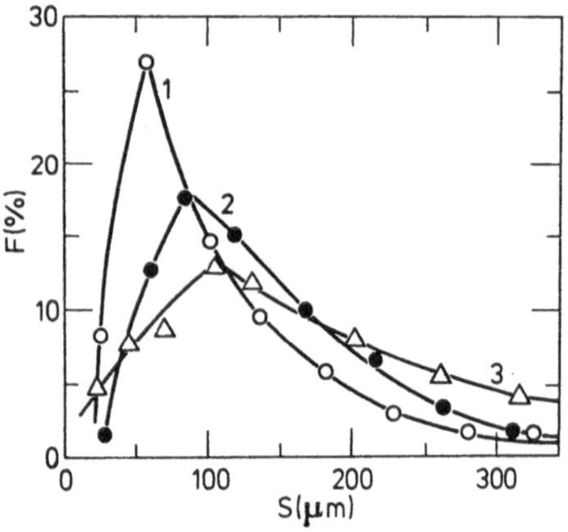

Fig. 26. Distribution frequency F of cell section radii s in foamed polymer based on poly(vinyl alcohol); (1) $\vartheta_p = 0.7$, $\bar{S} = 73\ \mu m$; (2) $\vartheta_p = 0.53$, $\bar{S} = 87\ \mu m$; (3) $\vartheta_p = 0.3$, $\bar{S} = 130\ \mu m$ [29]

samples had ϑ_p values ranging from 0.30 to 0.70. For the denser foam ($\vartheta_p = 0.70$) the distribution of s is Poissonian: a sharp peak is observed at relatively small s (curve 1 in Fig. 26). As the packing factor is decreased, the curves shift to larger values of s (curves 2 and 3, Fig. 26). The distribution pattern of s then resembles the regressive or the triangular distribution curves. The mean values $\bar{s}$ can vary up to factor 2 under these conditions.

On further decrease of the volumetric weight, the cell shape becomes too different from spherical, so that the Mihira's classification of cell size distribution and calculation procedures can be used but with caution [116-118].

9.4 Influence of Mechanical Deformation on the Cell Size Distribution

Next to nothing is known about the effect of mechanical loading on the cell size distribution or mean cell size. Such a lack of data is due to the difficulties attending all attempts to obtain such information.

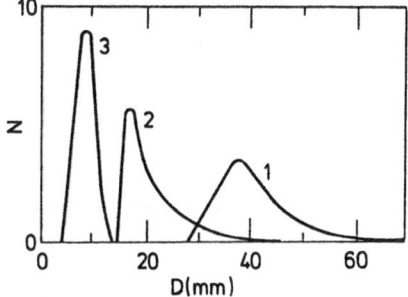

Fig. 27. Cell diameter D distribution (cell number N in relative units) of foamed polymer based on poly-(vinyl formal) at compression deformation ε (%): (1) 0.2, (2) 0.257, (3) 0.457 [119]

The radioactive-labeled liquid penetration method has been used by Lashnev [119] for the determination of the cell size distribution curves of wet poly(vinyl formal) foams for different strain parameters (Fig. 27). The results convincingly demonstrate the considerable structural changes occurring in polymer foams under compressive load. If we furthermore take into account the strong influence of the cell size on the technical characteristics of plastic foams, the importance of the pertinent data will become apparant. Since flexible plastic foams are often employed not only as resilient dampers but also as heat and sound insulating materials, to calculate the thermoconductivity coefficient of a foam structure, it is important to know not only the original structure but also size, porosity of cells in a plastic foam under mechanical stress [120-123].

The microscopic visual study can only cover a small portion of cells, namely, those located on the sample surface. Besides, the destruction of surface cells is too specific and does not fully reflect the general behavior and destruction of the bulk of the cells located inside the sample.

10 Surface Area of Foamed Polymers

10.1 Terminology

The term "surface area" (S) may be defined in different ways:
1) visible surface, e.g. the geometrical surface area of a foam section (S_{vis});
2) external surface (S_{ext}) of a plastic foam equal to the sum of the external areas of all cell walls constituting the external surface of material;
3) internal surface (S_{int}), i.e. the wall surface area of cells, micropores, cracks, fractures, etc.; and
4) total surface, S_Σ i.e. the sum of the external and the internal surface ($S_\Sigma = S_{ext} + S_{int}$) of a foam sample.

10.2 Specific Surface Area

The term "specific surface area" (S_{sp}) refers to the ratio of the total or only external foam surface to the volume V or mass (M_p) of the polymer phase:

$$S_{sp}^V = \frac{S_\Sigma}{V} \; ; \quad S_{sp}^M = \frac{S_\Sigma}{M_p} \tag{55}$$

where S_{sp}^V is the volumetric specific surface (cm^2/cm^3) and
$\quad$ S_{sp}^m the weight (mass) specific surface (cm^2/g).
Both parameters are related by

$$S_{sp}^M = S_{sp}^V/\gamma \tag{56}$$

where γ is the volumetric weight of the plastic foam (g/cm^3).

In analogy with Deryagin's definition of the active porosity [101], a kinetic specific surface was introduced (S_{Sp}^K):

$$S_{sp}^{KV} = \frac{S_d}{V_d}, \text{m}^2 \text{ m}^3 \qquad S_{sp}^{KM} = \frac{S_d}{m}, \text{m}^2/\text{g} \tag{57}$$

The kinetic specific surface is equal to the ratio of the surface area of the dispersant-conducting pores or cells (S_d) to the volume (V_d) or mass (m) of dry matter in porous material. On an anatomic scale no surface of any part of a solid body can ever be really planar. It usually contains cracks or fractures, many reaching deep into the material. Even minor cracks or wrinkles contribute to the surface area. The internal surface of porous bodies is usually several orders of magnitude larger than the external surface so that the total surface of such material is practically made up by the internal surface.

The internal surface area of plastic foams may be measured not only by direct physico chemical methods but also indirectly by morphological parameters of the structure. Thus, if we assume that the solid phase of the plastic foam contains no microcells and that the macrocells are spherical, the value S_{sp}^V will be given by

$$S_{sp}^V = \frac{6\pi \sum_i D_{p_i}^2 N_i}{\pi \sum_i D_{p_i} N_i} \cong \frac{6}{\gamma_p \bar{D}_p} \tag{58}$$

where γ_p is polymer phase density and
$\quad$ $\bar{D}_p$ mean probable cell diameter (see Sect. 8.1).
If the cell size distribution in the material is normal, then

$$S_{sp}^M = \frac{\alpha_S}{\alpha_V} \cdot \frac{\sigma_a^2 + \bar{D}_a}{3\bar{D}_a\sigma_a^2 - 2\bar{D}_a^2} \tag{59}$$

where $\bar{D}_a$ is mean arithmetic cell diameter (see Sect. 8.1) and
$\quad$ σ_a the standard deviation (root-mean-square deviation) α_S, α_V are the surface and volume aspect ratios of cells (see Eq. (28)).
If, on the other hand, the distribution follows a logarithmic law, then:

$$\log S_{sp}^m = \log \frac{\alpha_s}{\alpha_v}\gamma_p - \log \bar{D}_g - 5.76 \log^2 \sigma_g \tag{60}$$

where $\bar{D}_g, \sigma_g$ are the geometrical mean cell size and deviation, respectively.

As pointed out in Section 7, the plastio foam cells have a very complex shape, most frequently that of irregular polygons, which makes the use of the above formulas too cumbersome; nevertheless, the order of the specific surface value given by the equations is correct.

Vice versa, starting from the specific surface of a foamed polymer one can find the geometrical size of the constituent cells. Indeed, in those rare cases where the real cellular foam structure may be represented by sphere$^\lambda$ of the same size, the diameter of the spheres can be easily calculated from Eq. (58) [90]. In the case of more irregular structures, the diameter given by this formular is a "mean surface-volume diameter".

Using the concepts of active gas-filling and kinetic specific surface S_{sp}^K, it is possible to determine [101] the hydraulic pore or cell radius r_h as:

$$r_h = V_m/S_{sp}^K \tag{61}$$

r_h is equivalent to the average thickness of the liquid layer overlying the material particle surfaces and determines the mean diameter of cells which are fluid-permeable.

10.3 Roughness Factor

According to Briscall [102], the external surface area of a foam Section (S_{ext}) equals the product of the geometrical area of the section and a "roughness" factor R_B, i.e. $S_{ext} = S_{vis} \cdot R_B$ where:

$$R_B = \frac{S_{ext}}{S_{vis}} = \frac{S_{vis} + \Delta S_z}{S_{vis}} = 1 + \frac{\Delta S}{S_{vis}} \tag{62}$$

where S_{ext} is the true sectional surface area
S the geometrical section surface area and
ΔS the external surface area increment contributed by the surfaces of the dissected cells.
It is easy to fixd ΔS when the cells, chaotically located in the material bulk, are spherical and overlap nowhere; for a section diameter of D_i then ΔS is:

$$\Delta S_i = \frac{\pi}{2} D_i^2 \tag{63}$$

For the entire sectional area containing n section circumferences we obtain

$$\Delta S = \frac{\pi}{2} \sum_{i=1}^{n} D_i^2 \tag{64}$$

where $i = 1, 2 \dots n$.

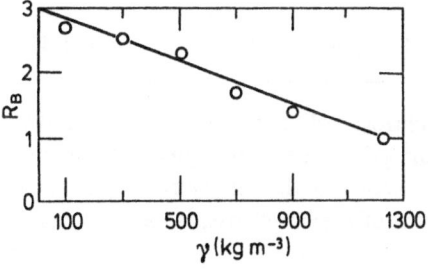

Fig. 28. Relationship between "roughness factor" R_B and volumetric weight γ of foamed polymer [102]

The calculation of R_B by counting the number of dissected cells and their diameters has shown that the factor R_B is inversely proportional to the volumetric weight of the plastic foam (Fig. 28) and, according to contact print data, does not exceed $R_B = 3$.

However, contact prints provide no information on the interior relief of a section so that there are very few results. Errors in the determination of R_B using this method are particularly tangible for lightweight plastic foams which incorporate a large proportion of overlapping cells and broken cell walls. The true value of R_B, all these factors considered, may be as high as $R_B = 10$ [103].

10.4 Specific Cell Volume

Fig. 18 allows to calculate by optical bar charts of plastic foams, the morphological term: effective specific pore (cell) volume, V_{eff}; it can be expressed by S_{sp}^M as follows [72]:

$$V_{eff} = S_{sp}^M \bar{D} = \frac{6 \sum_i v_i \mu_i^2}{\gamma_p \sum_i v_i \mu_i^3}$$

(65)

$$v_i = \frac{\Delta N_i}{N}; \qquad \mu_i = \frac{D_i}{\bar{D}}$$

where D_i is the i-th cell diameter
ΔN_i the number of cells with diameter D_i and
$\bar{D}$ the mean cell diameter.

It is seen from Eq. (65) that V_{eff} depends only on the cell size distribution in the material and that it should be known apart from the specific pore (cell) volume which is defined as

$$V_{sp} = \frac{1}{\gamma} - \frac{1}{\gamma_p}$$

(66)

where γ_p and γ are the polymer density and volumetric weigt of plastic foam, respectively.

Apparently, V_{sp} is controlled by the volumeric weight of the foam and is a characteristic of the gas phase content.

Using Eq. (65), V_{eff} has been calculated for different grade PPU-3 polyurethane foam samples from the optical bar chart data:

γ (kg/m³)	60	60	130	150	160	200	210	250	290	410	500
N		60.3	719	495	461	743	596	567	524	599	683	647
V_{eff} (cm³/g)		0.73	0.63	1.01	1.16	1.43	1.24	1.66	1.28	1.17	2.09	1.42

These results suggest than there is no straightforward relation between V_{eff} and γ, V_{eff} fluctuating about a mean value of 1.26 cm³/g. This means that (just as for microcells) the macrocell distribution pattern is independent of the volumetric weight of the foam and, judging from the asymmetrical form of the bar charts, is logaithmically-normal.

10.5 Foamed Polymers as High-Dispersive Systems

This author [87, 88, 104] and Fedossev and Litvinova [105] showed that foamed phenolic resins yielded an unexpectedly high S_{sp} value, ranging from 5 to 200 m²/g, which is 2 to 3 orders of magnitude higher than the calculated S_{sp} values obtained from visual (geometrical) measurements by multiplying the number of macrocells in a section by their spherical surface area. The reason for this became clear only after microcells had been discovered in oligomeric foams and found to constitute the bulk of all foam cells [7, 8].

Colloid chemistry classifies materials having S_{sp} of tens or hundreds of square meters per gram as disperse systems. More important is that plastic foams exhibit properties of disperse systems. For example, the thermooxidative stability of phenolic foams was lowered by several dozen degrees as S_{sp} was increased [88, 106].

In colloid chemistry all kinds of foams, both liquid and solid, are classified as low- or coarse-disperse systems, i.e. such in which the minimum pore diameter is not less than 1 mm. The grouping of plastic foams into coarse disperse systems has until recently, been justified, since the minimum size of the cells that could be observed did not exceed this value. However, microcells have been discovered in polymer foam structures, this requires a revision of the old concepts regarding the dispersity. The real minimum size of microcells is a few hundredth of a micron and their number is well above that of macrocells. These results support the conclusion that plastic foams belong to the group of finely dispersed or colloid systems [8].

10.6 Problems

The fact that foamed plastic can be classified among fine-dispersed systems is highly important since their structure may be studied not only from the conventional "polymer" point of view, but also by applying a different novel and promising approach — the physics and chemistry of disperse materials.

Such an approach involves considerable potential from the wealth of ideas and experience accumulated with "nonpolymer" disperse systems on the one hand, and the first results already obtained by applying this approach to foamed polymers on the other [87, 88, 104–106]. This approach opens a principally different way of control and presetting of the physicochemical properties of plastic foam products by varying one morphological parameter, the specific surface area; this may be done by regulating the size and number of microcells.

The relation between the specific surface and properties of foamed plastics is discussed in a monograph [88].

11 Cellular Structure Models and Calculation of Mechanical Properties of Foamed Polymers

11.1 Flexible Foamed Polymers

As noted above, the adequate simulation of the cellular structure of plastic foams is very important for the investigation of the relation between structure and properties of foamed plastics. A solution to this problem would allow to establish not only the general interrelation of the structural parameters and mechanical properties of plastic foams and gain an insight into their deformation mechanisms but also — even more important — would allow to work out a mathematical apparatus for the calculation of the quantitative relation between the factors under real service conditions.

Gent and Thomas [65, 76] were among the first to attempt a theoretical treatment of the straining of flexible gas-filled polymers for rubber foam samples. They proposed a structure model consisting of an arbitrary number of limear filaments joined at their ends into monolithic bundles. The mathematical treatment of a special case, rectangular cubic lattice, permitted only a semiquantitative evaluation for the mechanical behavior of flexible plastic foams.

Later, Polyakov and Tarakanov [77] modified the model by presenting if of a hexahedral cell having an initial curvature (eccentricity) near the rods disposed in two perpendicular directions. This model, which makes allowance for the initial anisotropy of the plastic foam, satisfactorily describes the elastic properties of flexible foams under considerable strains, but generally predicts unduly high values of the elastic properties of foamed plastics.

Dementiev and Tarakanov [78] calculated formulas for the strength and elastic properties of rigid and flexible oligomeric and thermosetting plastic foams, assuming the cells to be 14-hedrons (See Chapter 7.2). Their formulas are in fair agreement with the experimental results for both open- and closed-cell poly(vinyl chloride)- and polyurethane-based foams. This model characteristically provides for the strut deflection even under a small load deforming a sample in different directions relative to the foaming direction. The subcellular structure of this type is composed of rods joined together into squares. Adjacent squares contact each other only in perpendicular directions; in anisotropic foams the rhombic diagonals undergo elongation in the direction of expansion. It is assumed that the rods of the square cross section are secured at undeformable points so that the height is much less than the rod length. It is further assumed that under both small and heavy strain the rods have to resist lateral bending and that an energetically optimum form of the rod stability loss — cooperative buckling of the rods proceeding from a common point — occurs when the limit of compressive strength is uchieved.

Within such model, the polymer volumes enclosed in an elementary cell of an isotropic and anisotropic foam are respectively:

$$\frac{\gamma}{\gamma_p} = \frac{3\sqrt{2}\beta^2}{(2+\beta)^2} \tag{67}$$

and

$$\frac{\gamma}{\gamma_p} = \frac{k\beta^2(4k + 3k\beta + 2)}{\sqrt{1 - \frac{1}{2k^2}(2+\beta)(2+k\beta)^2}} \tag{68}$$

where β   is the ratio of rod width to the initial length of a longitudinal rod
     k   the ratio of the initial length of a longitudinal rod to the initial length of a transverse rod and
    γ, γ_p are plastic foam and polymer matrix densities, respectively.

From this model it follows, for example, that under small compressive loads the flexibility modulus E_f, the foam strain at compression ε_c, and the compressive strength σ_c are respectively:

$$E_f = E_p \frac{\gamma(2+\beta)}{18\gamma_p} \tag{69}$$

$$\varepsilon_c = \frac{\pi^2 \beta^2}{8(2 + \beta)} \tag{70}$$

$$\sigma_c = \left(\frac{\gamma}{\gamma_p}\right)^2 \frac{\pi E_p (2 + \beta)^2 \cos 45°}{432} \tag{71}$$

where E_p is the polymer matrix elasticity modulus.

Table 4 compiles the experimental and calculated data on some isotropic and anisotropic plasic foams obtained from Eqs. (69)–(71).

Table 4. Comparison of Calculated and Experimental Mechanical Characteristics of Certain Plastic Foams [78]

Foam	γ (kg/m³)	E_f (kgf/cm²)		ε_c (%)		σ_c (kgf/cm²)	
		exper.	calc.	exper.	calc.	exper.	calc.
Poly(vinyl chloride) foam	76	0.21	0.38	9.0	4.6	0.019	0.021
Polyethylene foam	66	—	—	4.9	3.7	0.83	0.49
Polyurethane foam	50	105	130	3.0	2.7	3.5	3.9

11.2 Rigid Foamed Polymers

According to Dementiev and Tarakanov [124], under mechanical loading, the rigidity of foams always changes abruptly when a certain stress σ_c is reached (the nominal critical compressive strength) when the struts are bent and crumpled. Dependisg on the shape of the compression curves and the failure mechanism, all rigid foamed plastics may be divided into five classes (Fig. 29). In the case of lightweight plastic foams the cellular structure fails, due to the loss of strut stability at σ_c. In large-pore rigid "Penoisovinyl" PVC foams (Fig. 29a), at first the struts of the weakest layer, approximately one cell deep, fail abruptly and the load drops considerably at the same time. Then the struts of each consecutive adjacent layer are destroyed in turn. In *small-pore* lightweight plastic foams (Fig. 29b) the process also begins with the crumpling of the struts of the weakest transverse layer, which is, however, not one but several cells deep. This layer then expands due to the consecutive crumpling of the adjacent layers.

In the compression diagram this stage is revealed either as a plateau or a slight increase of load on continued strain. After release of the load and prolonged rehabilitation at elevated temperature the samples normally acquire their original dimensions. In large-pore *semirgid* plastic foams the struts of the weakest layer, about one cell deep, experience bending on attainment of σ_c (Fig. 29c); in the following stages the struts of each consecutive adjacent layer are also bent. In the compression diagram this stage is represented by either a plateau portion or a minor increase of load when the compression is continued. After release of the load and prolonged rehabilitation at room temperature the samples assume their original

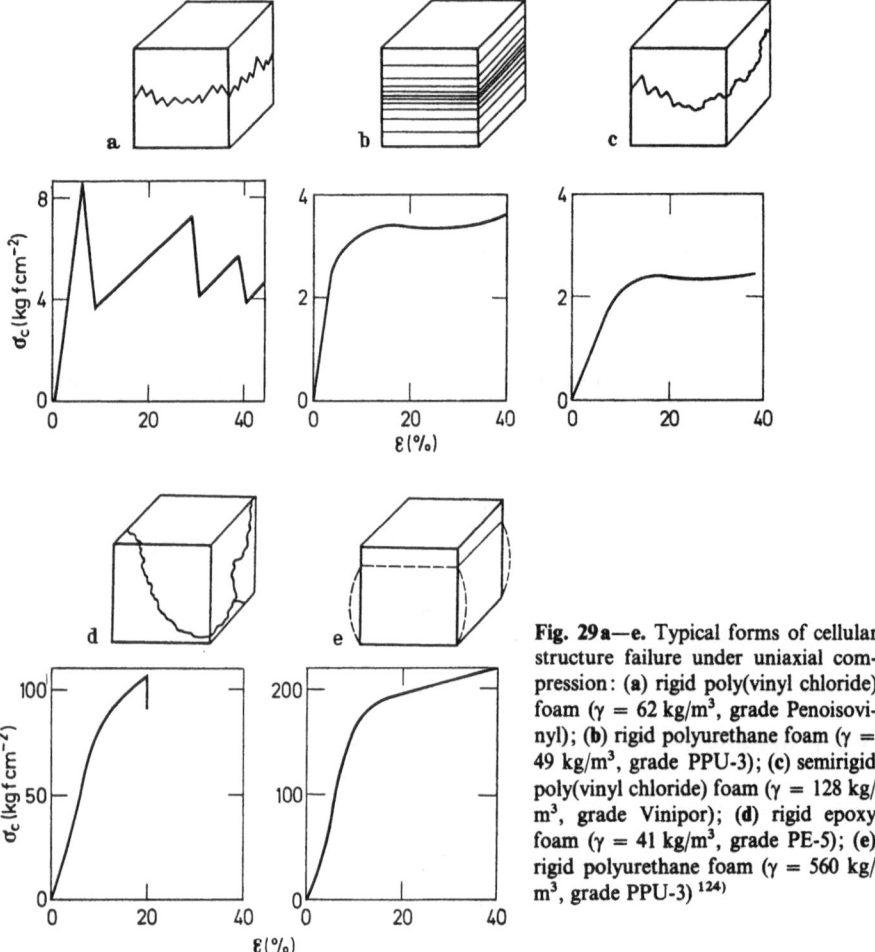

Fig. 29a—e. Typical forms of cellular structure failure under uniaxial compression: (**a**) rigid poly(vinyl chloride) foam ($\gamma = 62$ kg/m³, grade Penoisovinyl); (**b**) rigid polyurethane foam ($\gamma = 49$ kg/m³, grade PPU-3); (**c**) semirigid poly(vinyl chloride) foam ($\gamma = 128$ kg/m³, grade Vinipor); (**d**) rigid epoxy foam ($\gamma = 41$ kg/m³, grade PE-5); (**e**) rigid polyurethane foam ($\gamma = 560$ kg/m³, grade PPU-3) [124]

dimensions. In all cases (a–c) the transverse dimensions of the samples remained practically unchanged even under compressive strains of up to 50%.

Denser foams behave differently in compression tests. The brittle failure of samples involves cracking along the inclined and longitudinal planes (Fig. 29d). When compressive load is applied to a material of high apparent density, such as polyurethane foam (Fig. 29e), the load will go on increasing continuously even after σ_c has been attained. At the same time, the cross-section of the sample will increase and the sample assume a barrel-like shape. The foam cells will begin to crumple at the same time.

Similarly, the analysis of compression diagrams shows that for lightweight rigid plastic foams there are usually three well-defined regions of the $\varepsilon = f(\sigma)$ curve: 1) an initial region describing a sharp increase of deformation; 2) a plateau or a minor increase of strain; and 3) a sharp decrease of resistance to deformation. The experimental data seem to support the following interpretation: The initial portion corresponds to the compression and bending of the GSE struts and walls until they totally lose resistance (buckle); the second portion describes buckling and failure or bending of the struts, due to forced elastic strain which results in a sharp load drop in the former case or a minor increase of load in the latter; the third portion reflects the final crumpling of the buckled struts and a gradual compaction of the polymer. The behavior of denser foams under compression is closer to that of monolithic plastic materials [120-124].

In the work of Dobrovolsky et al. [125] the behavior of the cellular structure of rigid plastic poly(vinyl chloride) foams (grade PVKh-1) under monotonously increased tensile load has been simulated by that of a rod structure. A cellular structure was modeled by a continuum corn posed of a plurality of interconnected rod elements which are arbitrarily oriented with respect to the applied tensile load P. In the simplest case, a load-bearing element of such a framework consists of at least two identically sized rods (Fig. 30). The geometrical orientation of the elements relative to the applied force P is characterized by the rod inclination angle (α) relative to the plane perpendicular to the direction of the force P. On the "macroscale", the rods with different angles α_i are located at random and are so numerous that their distribution is continuous. For isotropic plastic foams it may be assumed that the distribution of α is uniform in the range from 0 to $\pi/2$. Concequently, the function $F(\alpha_i)$, describing the probability of finding rods whose angle of inclination is below a preset α_i, is $2\alpha_i/\pi$.

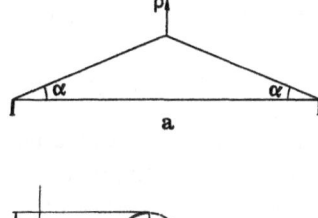

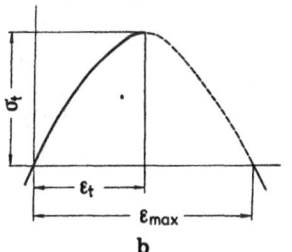

Fig. 30a and b. Rod model of a cellular structure (a) and tension-failure diagram (b) of this model for rigid foamed polymers [125]

When a monotonously increasing load P is applied to such a load-bearing element, its failure begins with those rods which are directed closer to the direction of the applied load. Obviously, the stresses in whole rods will be the same and in failed rods will be zero, and the model is deformed on account of the change in rod orientation.

If the plastic foam failure in fact occurs in this manner, it may be assumed that each value of strain (ε_i) corresponds to some actual cross-sectional area $S_a(\varepsilon_i)$ which is different from the original area S_0 because some rods have been destroyed by the force P. Since the active cross-section which experiences the load is proportional to the number of unfailed joints, then:

$$S_a(\varepsilon_i) = S_0 F(\alpha_i) = S_0 \frac{2\alpha_i}{\pi} \tag{72}$$

If we assume that the ε_i versus α_i relationship is linear, then

$$\varepsilon_i = \varepsilon_{max} - \frac{2\varepsilon_{max}\alpha_i}{\pi} \tag{73}$$

$$\alpha_i = \frac{\pi}{2}\left(1 - \frac{\varepsilon_i}{\varepsilon_{max}}\right)$$
(74)

Hence

$$S_0(\varepsilon_i) = S_0\left(1 - \frac{\varepsilon_i}{\varepsilon_{max}}\right)$$
(75)

where ε_{max} is the maximum deformation at complete failure (Fig. 30).
At equilibrium of the destruction, the force P is equal to

$$P = E_0\varepsilon_i S_a(\varepsilon_i)$$
(76)

where E_0 is the elasticity modulus of rods.

Using now stresses instead of forces and expressing the elasticity moduli of the model (E) in terms of E_0, and the system and rod densities in terms ou C_p, we may write:

$$\sigma_i = \omega E_0\varepsilon_i\left(1 - \frac{\varepsilon_c}{\varepsilon_t}\right)\frac{\gamma}{\gamma_p}$$
(77)

where ω is a numerical parameter accounting for the system configuration which was calculated as in [125a].

From Eq. (77) it follows that the deformation at ultimate elongation (ε_t) occurring at the maximum applied load is 0.5 ε_{max} (Fig. 30).

Comparing the calculated and experimental stress-strain diagrams for real plastic foams, we will take E_0 in Eq. (77) as the elasticity modulus of the polymer base; γ and γ_p are foam and polymer base densities (volumetric weights), respectively; ε_{max} is the maximum deformation equal to double the ultimate elongation (ε_t) of the plastic foam. With an accuracy of up to 8%, Eq. (77) allows to approximate the tensile stress σ_t for rigid poly(vinyl chloride) and polystyrene foams. For example, for polysterene foam PSB ($\gamma = 20$ kg/m³, $\gamma_p = 1050$ kg/m³, $E_0 = 3.10^4$ kgf/cm², $\omega = 0.33$) predicted σ_t^p values are the following:

ε_t (%)	0.4	0.8	1.2
σ_t^p (kgf/cm²)	0.85	1.35	1.55
σ_t^m (kgf/cm²)	0.80	1.41	1.59

Eq. (77) may be valid for other rigid foams as well, since the model underlying the calculation procedure encompasses morphologies of a large enough range of real cellular polymers.

12 Cellular Structure and Thermoconductivity of Foamed Polymers

The thermoconductivity (λ) of plastic foams is generally composed of the thermoconductivity of the solid phase (λ_s), the gas (λ_g), and the convective (λ_c) and radiative (λ_c) and radiative (λ_r) components [126-128]:

$$\lambda = \lambda_s + \lambda_g + \lambda_c + \lambda_r.$$
(78)

When plastic foam is intended for use as a heat insulating material it is desirable to reduce the contribution of each of the above components to the total value λ.

12.1 Thermoconductivity of the Polymer Phase

The contribution of λ_s is very small for two reasons. Eirst, the thermoconductivity of the polymer phase is never high and usually between 0.1–0.3 kcal/m × h °C [129]. Second, the polymer phase (the GSE walls and struts) occupies but a small fraction of the space of material. For example, for a volumetric weigt of 32 kg/m³ the polymer occupies only 3% of volume which gives for polystyrene λ_s = 0.00248 and for polyurethane γ_s = 0.00496 kcal/m × h °C. However, a further decrease of λ by reducing the solid phase volume ratio is not always possible (because not all polymers can be produced with a very low density), feasible (for economic or technological reasons) or desirable (the smaller the volumetric weight, the poorer the strength of foamed plastics). Harding [66] found that for plastic foams with a markedly anisotropic structure the contribution of the solid phase conductivity to the total heat transfer increases with rising cell aspect ratio ($\alpha = D^{\parallel}/D^{\perp}$) (Fig. 31).

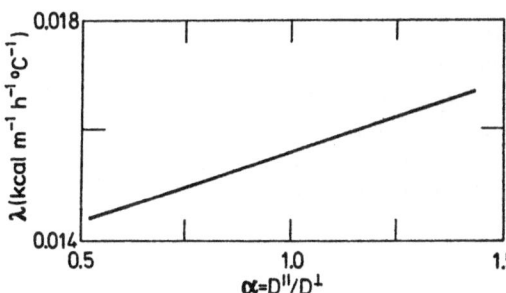

Fig. 31. Thermoconductivity λ of $CFCl_3$-foamed polyurethane as a function of cell orientation $\alpha = D^{\parallel}/D^{\perp}$ relative to heat flux; $D^{\parallel}$ and $D^{\perp}$ are the mean cell diameters parallel and perpendicular to the heat flux respectively [66]

12.2 Thermoconductivity of the Gas Phase

The gas in the cells is responsible for the major part of the heat transfer, since the volume ratio of a gas in a foam usually exceeds 90%. Since the number of blowing agents that can be used to produce plastic foams is practically unlimited, the component λ_g may be varied in a very broad range [130–133]. Evidently, the gas should have a low λ_g value, i.e. a high molecular mass (Fig. 32); this is, however, not always possible or desirable in practice, for economic or technical reasons [134, 135].

Afer the foaming is complete, the blowing agent is replaced with air due to diffusion. The moment at which the diffusion equilibrium is achieved is influenced by the following factors:

1) the chemical nature of the polymer and the gas, and especially the permeability of the cell walls of the polymer to a given gas or air;

2) cell size; a large number of small cells will resist gas exchange better than a small number of large cells;

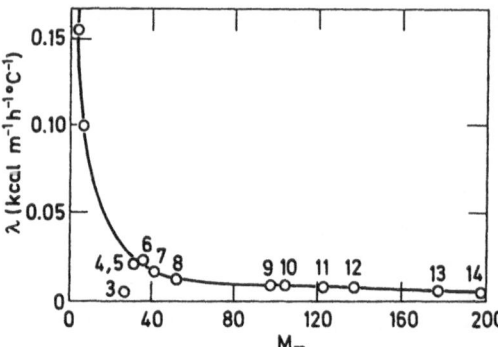

Fig. 32. Thermoconductivity λ of various blowing agents as a function of their molecular mass M_m at 15.6 °C: (1) H_2, (2) He, (3) H_2O, (4) N_2, (5) air, (6) O_2, (7) CO_2, (8) CH_3Cl, (9) F-22, (10) F-21, (11) F-12, (12) F-11, (13) F-114, (14) F-115 [135]

3) degree of cell closeness and whether or not there is a compacted skin layer (Fig. 33a);

4) shape of the foam article: first in the surface zones and only later in the deeper zones is gas replaced by ambient atmosphere; therefore, the thicker the sample, the more time will the equilibrium take to be established;

5) storage conditions (ambient temperature and humidity).

From all these factors the time necessary fo a diffusion equilibrium may be a few years, as shown by Küster [136] (Fig. 33b). Carbon dioxide which is often used as a blowing agent has a lower heat conductivity (0.014 kcal/m × h °C) than air (0.023 kcal/m × h °C at 20 °C). Thus, CO_2 would seem tobe of advantage in the manufacture of high-quality heat insulating foams. However, most polymers are much more permeable to CO_2 than to air, thus the former tends to be rapidly displaced by the latter. Hence, despite its attractively low thermoconductivity,

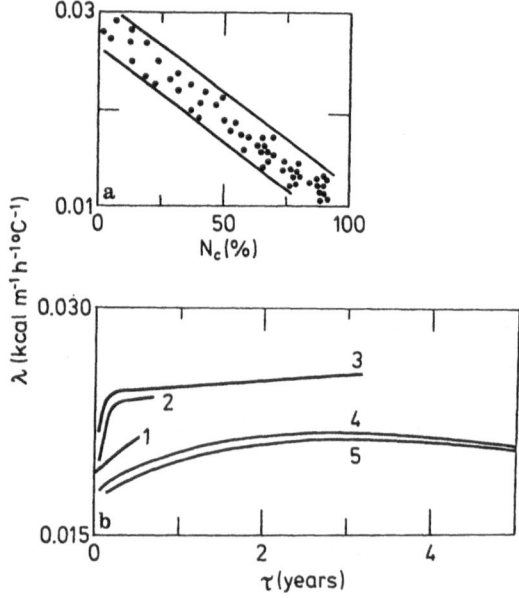

Fig. 33a and b. Variations of thermoconductivity λ of $CFCl_3$-foamed polyurethane foam: (a) as a function of the number of closed cells N_c (%) (b) as a function of storage time τ and temperature: (1) $\gamma = 45$ kg/m³, 20 °C; (2) $\gamma = 45$ kg/m³, 50 °C; (3) $\gamma = 45$ kg/m³, 80 °C; (4) $\gamma = 32$ kg/m³, 24 °C; (5) $\gamma = 30$ kg/m³, 60 °C [136]

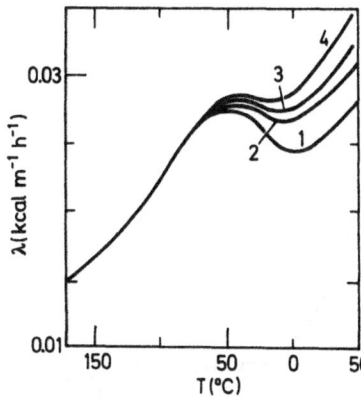

Fig. 34. Thermoconductivity λ of freon-foamed poly-styrene as a function of storage temperature for different storage periods: (1) 0.5; (2) 1; (3) 1.5; (4) 2 years [142]

carbon dioxide does not provide an accordingly low thermoconductivity of the product [137-140].

In contrast, a much longer time is needed for the diffusion equilibrium to be established in plastic foams expanded with freons, since CCl_3F features a consider-ably lower diffusivity through cell walls than air. As freon vapors feature a very low thermoconducivity, we clearly see why they should preferably be used as blowing agents for heat insulating plastic foams [141].

Zehender [142] has found that the thermoconductivity curves for freon-foamed plastic have the shape shown in Fig. 34. At below -60 °C and above $+10$ °C, λ increases nearly in proportion with temperature. However, there are an intermediate maximum and a minimum of λ in the range between -60 and -24 °C. The obvious reason is: trichlorofluoromethane vapor condensates in this temperature range. When CCl_3F is replaced by air in the cells, the bend of the curve in the temperature range from -60 to -10 °C gradually becomes less pronounced. However, even two years later the presence of freon vapors manifests itself in the temperature dependence of the λ-value, indicating a sufficiently low rate for the diffusion equilibrium in the "freon-polymer-air" system. The replacement of the blowing agent with air takes place the slower, the more closed cells there are in the material and the denser the skin layer of the foam [136, 143].

12.3 Radiative Thermoconductivity

The contribution of the radiative heat exchange in polymer foams depends, naturally, on the radiative capacity of the starting polymer material (about 0.9); also, according to the Stefan-Boltzmann law, λ_r is influenced considerably by the temper-ature gradient ΔT prevailing between the sample surfaces (Fig. 35). The contribution of λ_r to the total value λ is of secondary importance although it may be rather high for lightweight foams. This can be inferred, for instance, from the light transmission data on polystyrene foams of different densities (without allowance for light scattering) (Fig. 36).

It is quite impossible to measure the share of λ_r in λ by direct measurements, since the obtained relationship is valid only for a specific temperature (in the presented case, that of an incandescent lamp was 1400 °C [126].

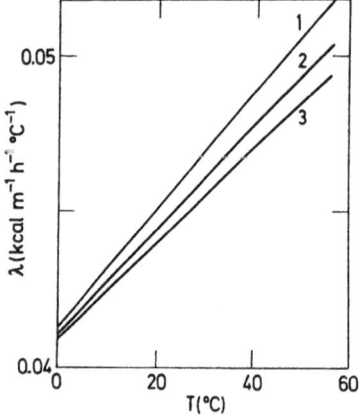

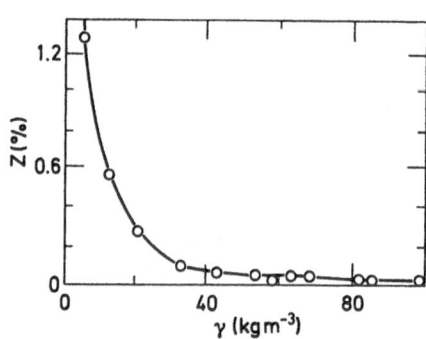

Fig. 35. Thermoconductivity λ of polystyrene foam ($\gamma = 5$ kg/m³) as a function of temperature under different temperature gradient ΔT between sample surfaces; ΔT (°C): (1) 4, (2) 10, (3) 18 [144]

Fig. 36. Transparency Z of polystyrene foam as a function of volumetric weight γ [126]

The share of the radiative heat transfer in the total heat exchange largely depends on the cell size. Since heat transfer by radiation is established through a wall separating any two cells oriented in its direction, a foam consisting of a *large* number of *small* cells will (under similar conditions) resist the radiative heat transfer more effectively than a material comprising a *smaller* number of *larger* cells. This explains why an increase of the cell size has a stronger effect on the thermoconductivity than a rise in the volumetric weight [133]. When the sample thickness is raised, λ_r will increase even if the cell wall thickness were decreased. However, for a sufficiently large volumetric weight of the material (above 100 kg/m³) the effect of the sample thickness becomes negligibly small. The contribution of λ_r is the smaller, the greater the number of cells in a foam and the smaller they are in size.

12.4 Convective Thermoconductivity

Convective fluxes arise in gas-filled cells whenever the cell size exceeds a threshold value. Some investigators estimated the theshold size between 2 and 5 mm [131] [134] [144]. A convincing proof of the convection in the cells of foamed plastics was obtained from polystryrene foam samples of the same volumetric weight but of different cell sizes [140]. The thermoconductivity was measured stationary in two positions: first, the heater was placed underneath the sample and a cooler above it, then their positions were reversed. Thus, the heat flux first propagated against the gravitation which increased the intensity of convective mixing. In the second case the direction of heat flux was the same as that of gravitation. The deviations between the λ values for the two arrangements of heater became noticeable at a cell diameter of above 4 mm.

However, since most of the plastic foams now produced commercially have a cell size below this value, the contribution of the convective heat exchange to the total heat

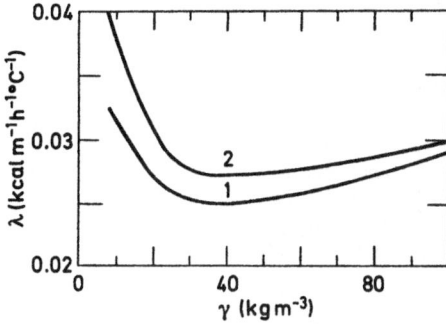

Fig. 37. Thermoconductivity λ of polystyrene foam as a function of volumetric weight γ at different sample temperatures (°C): (1) 0, (2) 40 [140]

Table 5. Relative Shares (%) of the Components in the Thermo-Conductivity of Polystyrene Foam [126,145]

Component of λ	Volumetric weight γ (kg/m³)	
	25 (at 10 °C)	32 (at 22 °C)
Solid phase λ_s	7	6
Gas phase λ_g	63	74
Convective λ_c	4	0
Radiative λ_r	26	20

transfer is very small and becomes detectable only in very lightweight foams (Fig. 37).

The percentage of cells having a size exceeding 3 mm is large only for the most lightweight plastic foams (see Table 5); so the contribution of λ_c to λ is very small [145].

12.5 Influence of Different Factors

Thus, the contributions of the components λ_s, λ_r and λ_c being less important, the thermoconductivity of plastics foams, with the exception of the most lightweight ones, is largely determined by the gas phase composition.

The minimum in the $\lambda = f(\gamma)$ curve (Fig. 37) may be attributed to different heat conductivity mechanisms, depending on the cell size in the foam. Thus, in the region of small γ where the content of the solid phase is insignificant, the conditions are favorable for the radiative heat exchange. Vice versa, at higher densities, this component of heat exchange decreases since the number and wall thick ness of cells which act as "heat-insulating screens" become higher. At very low densities, with cell diameter being as large as a few millimeters, the convective heat transfer mechanism begins to manifest itself. As γ increases, the resultant rise in the proportion of the solid phase causes an increased heat flux through the polymer and decreased heat absorption, reflection and dissipation [146].

The thermoconductivity of a plastic foam sample also depends on its thickness, but varies not in direct proportion with it (Fig. 38). The particular sample thickness and

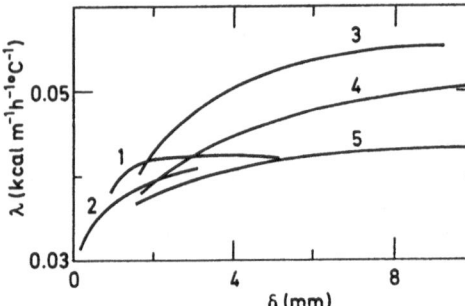

Fig. 38. Thermoconductivity λ of polystyrene foam as a function of sample thickness δ at different volumetric weight and temperature: (1) $\gamma = 10$ kg/m³, 30 °C; (2) $\gamma = 21$ kg/m³, 25 °C; (3) $\gamma = 20$ kg/m³, 15 °C; (4) $\gamma = 15$ kg/m³, 15 °C; (5) $\gamma = 29$ kg/m³, 15 °C [136]

configuration at which their effect on λ are minimal largely depend on the measurement techniques and instruments.

Another factor augmectting the heat conductivuty of plastic foams under conditions is the absorbed moisture. For example, for CCl₃F-foamed polyurethane at 25 °C and a relative humidity of 65%, the ambient moisture diffusion rate is 10–20 g/m² for 24 h. Especially strong is the effect of moisture on heat conductivity if the temperature differential across the sample is considerable. For example, in plastic foams used in cryogenic technology, the inner layers are exposed to low temperatures; the water vapor first condenses and is then convected into ice. Since the thermoconductivity of water and ice are 0.5 and 1.5 kcal/m × h °C, respectively, even minor amounts have a considerable detrimental effect of the heat insulating capacity of a foam material [147].

Gas and moisture diffusion may be prevented by the application onto the material surface of a tin, foil or synthetic coating; however, this is not always feasible technically or economically. For example, coating of a plastic foam with a 3 mm thick ABS-plastic film stabilizes the gas composition inside the foam for several years [136, 148, 149].

13 Conclusion

A principally novel approach to the investigation of plastic foam structures was recently outlined in polymer science. The approach is characterized by extensive use of methodological and factual data obtained from modern physical and physicochemical methods of structure investigations as by a more rigorous mathematical treatment. These investigations have resulted both in minute details and the general nature of the spatial morphology of polymer foams. They have fundamental morphological parameters and yielded to pass from the qualitative to the quantitative evaluation of these parameters on foam properties.

In order to gain a more profound knowledge, the first problem must be a wider use of the concepts of polymer physics and physicochemistry. This would enable an evaluation of the specific polymeric features of plastic foam morphology, primarily the different structures and types of the submolecular organization of cell walls and struts, and thus allow to understand their effect on the microstructure of the foam density and the degree of orientation of foam cells.

A more detailed study of foam dispersity and its relation with other morphological parameters seems to be no less important.

The next stage in such complex approach must be an investigation of polymer foams in the physics and mechanics of disperse systems [150, 151].

However, the morphological studies should not overshadow the main task, the investigation of the relation between morphology and properties of polymer foams. Indeed, all the morphological information is not valuable per se but is helpful towards understanding and, if possible, predicting the technical specifications and behavior of plastic foams under various external conditions, both in their manufacture and in applications [152-156].

These problems and ideas relate rightly to the physical aspect of polymer science. Other important scientific problems lie in the physicochemical direction — we mean the interpretation and prediction of the morphology of a given material from the chemical structure of the polymer matrix, material composition, expansion process and secondary processing conditions. To approach these problems, not only accumulation and systematization of formal graphical data, which was the object of this review would be required, but also ideas and methods from other sciences: colloid chemistry, physicochemical mechanics, polymer rheology, theory of the processes of manufacture and apparatus, etc. All we can do today within the physicochemical approach is to formulate the problem itself and the ensuing special cases of each particular polymer type, expansion process, operating conditions relating to apparatus, etc. We are still unable to solve these problems which will be a matter of the future.

Acknowledgement. The author is very grateful to Drs. G. Henrici-Olivé and S. Olivé (Monsanto Textiles Company, Pensacola, Florida, USA) for the very useful and numerous corrections and discussions of this paper.

14 References

1. Throne, J. L.: Proc. Intern. Conf. Polymer Processing, MIT-Press, Cambrige, USA 1979, pp. 77
2. Radushkevich, L. V.: Methods of Structural Investigations of Finely Dispersed and Porous Bodies, Moscow, AN SSSR Publishers 1958 pp. 281; Fundamental Problems in Physical Adsorption Theory, AN SSSR Publishers 1959, pp. 270 (Russ.)
3. Scheidegger, A. E.: Adv. Hydrosci. *1*, 161 (1965)
4. Nikolaevsky, V. N.: Izv. AN SSSR, ser. Mashinostroeniye *5*, 40 (1960) (Russ.)
5. Bernal, J.: Growth of Crystals. London: Interscience 1965
6. Minkovsky, H.: Gesammelte Abhandlungen *3*, 3 (1911)
7. Shutov, F. A., Chaikin, I. I.: Intern. Conf. Cellular and Noncellular Polyurethane, Strasbourg, France 1980, pp. 117
8. Shutov, F. A.: Adv. Polym. Sci. *39*, 1 (1981)
9. Dubinin, M. M.: Theory of Capillarity. Leningrad: Khimia 1980 (Russ.)
10. Ventzel, E. S.: Theory of Probability. Moscow: Nauka Press 1965 (Russ.)
11. Radushkevich, L. V.: Zh. Fiz. Khimii *40*, 965 (1966) (Russ.)
12. Giddings, J. C.: Dynamics of Chromatography, Part 1. New York: Marcell Dekker 1965
13. Berlin, A. A., Chaikin, I. I., Shutov, F. A.: Plast. Massy *9*, 42 (1981) (Russ.)
14. Bearman, R. J., Kirkwood, J. G.: J. Chem. Phys. *18*, 136 (1958)
15. Saffman, P. G.: Proc. Roy. Soc. *245A*, 312 (1958)
16. Bokiy, G. V.: Crystal Chemistry, Moscow: Nauka Press 1960, pp. 132 (Russ.)
17. Herdan, G.: Small Particle Statistics, London: Academic Press 1960
18. Smith, W. O., Foote, P. D.: Phys. Rev. *34*, 1271 (1929)

19. Hrubešek, J.: Koll. Beih. 53, 385 (1941)
20. Menke, H.: Phys. Z. 33, 593 (1932)
21. Prins, G. D.: Naturwissenschaften 19, 435 (1931)
22. Scott, G. D.: Nature 188, 908 (1960)
23. Shternberg, A. A.: Crystal Growth, Moscow: Nauka Press 1965, pp. 179 (Russ.)
24. Beneati, R. F., Brusilov, C. B.: Amer. Inst. Chem. Eng. J. 8, 359 (1962)
25. Kumpf, H.: Chem. Ind. Techn. 30, 144 (1958)
26. McGlory, R. K.: J. Amer. Ceram. Soc. 44, 513 (1961)
27. Krumbein, W. C.: J. Geology 43, 482 (1935)
28. Lissant, K. J.: J. Coll. Interface Sci. 22, 462 (1966)
29. Mihira, K., Ohsawa, T., Makayama, A.: Koll. Z. 222, 135 (1968)
30. Pöschl, T.: Z. Metallkunde 35, 25 (1943)
31. Flint, E. E.: Cristallography. Moscow: Nauka Press 1963 (Russ.)
32. Furnas, C. E.: Ind. Eng. Chem. 23, 1052 (1951)
33. Horaček, H.: Agnew. Makromol. Chem. 12, No. 80, 105 (1970)
34. Blair, E. A.: Conf. Cellular Plastics, Nattick, USA 1967
35. Jones, R. E., Fissman, G.: J. Cellular Plastics 1, 48 (1965)
36. Romanenkov, I. G.: Physico-Mechanical Properties of Foamed Plastics. Moscow: Standartgiz Press 1970, pp. 32 (Russ.)
37. Saunders, J. H., Frisch, K. C.: Chemistry of Polyurethanes. N.Y., London: Interscience Publ. 1964
38. Knox, R. E.: ASHRAE J. 10, 43 (1962)
39. Benning, C. J.: Plastic Foams. New York: John Wiley 1969, part II, pp. 10
40. Kafengauz, A. P., Kafengauz, I. M.: Plastich. Massy 9, 40 (1965) (Russ.)
41. Blair, E. A.: Structur of Plastic Foams, in: Resinography of Cellular Plastics. Philadelphia: ASTM 1967, ASTM Special Publ. No. 414, pp. 84
42. Cooper, A.: Plast. Inst. Trans. 26, 299 (1958)
43. Green, H. C.: Reticulated Foams, in: Encyclop. Polym. Sci. Technol. 12, 102 (1970)
44. Romanenkov, I. G.: Plastich. Massy 3, 51 (1967) (Russ.)
45. Thomas, C. R.: Brit. Plast. 9, 552 (1965)
46. Berlin, A. A., Shutov, F. A.: Chemistry and Technology of Foamed High Polymers. Moscow: Nauka Press 1980 (Russ.)
47. Mason, G. J.: J. Coll. Interface Sci. 35, 274 (1971)
48. Götse, H.: Schaumkunststoffe, Zirndorf: 1964, pp. 42
49. Moiseev, A. A., Durasova, T. F.: Foamed Plastics. Moscow: Oborongiz Press 1963, pp. 19 (Russ.)
50. Godilo, P. V., Paturoev, V. V., Romanenkov, I. G.: Non-Pressure-Molded Plastic Foams as a Structural Material. Moscow: Stroyizdat Press 1969, pp. 103 (Russ.)
51. Moiseev, A. A., Durasova, T. F.: Khim. Promyshl. 3, 16 (1967) (Russ.)
52. Valgin, V. D., Lebedev, V. S., Vasilieva, E. A.: Plastich. Massy 4, 37 (1967) (Russ.)
53. Murashev, Yu. S., Valgin, V. D.: Plastich. Massy 3, 22 (1968) (Russ.)
54. Stepanenko, V. V.: Candidate Thesis, Moscow: TsNII Stroitelnych Konstruktsiy Publication 1973 (Russ.)
55. Topper, L.: Ind. Eng. Chem. 47, 1377 (1955)
56. Gorring, R. L., Churchill, S. W.: Chem. Eng. Progr. 7, 53 (1961)
57. Doherty, D. J., Hurd, R., Lester, G. R.: Chem. Ind. 28, 1340 (1962)
58. Matonis, V. A.: SPEJ 20, 1024 (1964)
59. Gioumoussi, G.: J. Cell. Plast. 1, 224 (1965)
60. Ko, W. L.: J. Cell. Plast. 1, 224 (1965)
61. Harding, R. H.: IEC Process Develop. 2, 117 (1964)
62. Harper, J. C., El Sahrigi, A. F.: IEC Fundam. 4, 318 (1964)
63. Guenther, F. O.: SPE Trams. 2, 243 (1962)
64. Dawson, J. R., Shortall, J. B.: Europ. J. Cell. Plast. 2, 12 (1980)
65. Gent, A. N., Thomas, A. G.: Rubber Chem. Techn. 36, 597 (1963)
66. Harding, R. H.: Moisture Sorption of Foamed Plastics, in: Resinography of Cellular Plastics. Philadiphia: AST 1967, ASTM Special Publ. No. 414, pp. 3
67. Giaffria, R.: J. Polym. Sci. 60, 91 (1962)
68. Giaffria, R.: Mod. Plast. 36, 150 (1959)

69. Harding, R. H.: J. Cell. Plast. *1*, 81, 385 (1965); Effect of Cell Geometry of Foam Performance, in: Plastic Foams. Frisch, K. C., Saunders, J. H. (Eds.) New York: Marcell Dekker 1972, part II, pp. 831
70. Eisenklam, P.: Porous Masses. London: Butterworth 1956, Vol. 2, pp. 28
71. Vatuner, L. M., Pozin, M. E.: Mathematical Methods in Chemical Technology. Moscow: Khimiya Press 1972, pp. 387 (Russ.)
72. Gregg, S. J., Sing, K. S. W.: Adsorption, Surface Area and Porosity. London-New York: Academic Press 1967
73. Heywood, H.: Trans. Inst. Chem. Eng., Suppl. *1947*, 101
74. Karapetyan, O. O.: Candidacy Thesis. Leningrad: LISI Publication 1972
75. Karapetyan, O. O., Potapov, A. I., Gumina, I. I.: Stroit. Mekh. i Inzhen. Konstruktsii *1*, 15 (1969) (Russ.)
75a. Olivé, G., Olivé, S.: Priv. Comm. 1981
76. Gent, A. N., Thomas, A. G.: J. Appl. Polym. Sci. *1*, 107, 354 (1959)
77. Polyakov, Yu. N., Tarakanov, O. G.: Macrostructure Model for Flexible Polyurethane Foams, in: Mechanics of Fibrous and Porous Systems, Riga: Zinatne Press 1967, pp. 323 (Russ.)
78. Dementiev, A. G., Tarakanov, O. G.: Mekhanika Polimerov No 4, 594; *5*, 859 (*1970*); *1*, 45 (*1973*); *4*, 658 (1978) (Russ.)
79. Fehn, G. M.: J. Cell. Plast. *3*, 456 (1967)
80. Aleksandrov, A. Ya., Borodin, M. Ya., Pavlov, V. V.: Plastic Foam Insulated Structures. Moscow: Mashinostroyeniye Press 1972, pp. 10 (Russ.)
81. White, P. L., Van Vlask, L. H.: Metallography *3*, 241 (1970)
82. Aboav, D. A.: Metallography *5*, 251 (1972)
83. Talalay, J. A.: Ind. Eng. Chem. *46*, 1530 (1954)
84. Conant, F. S., Wohler, L. A.: India Rubber World *2*, 179 (1949)
85. Brumfield, H. L., Estill, W. B.: J. Cell. Plast. *7*, 212 (1971)
86. Spektor, F. O., Shutov, F. A.: Morphology of Phenolic Foams, in: Physics. Leningrad: LISI Publication 1970, pp. 11 (Russ.)
87. Shutov, F. A.: Candidate Thesis. Moscow: IKhF AN SSSR Publication 1971 (Russ.)
88. Berlin, A. A., Shutov, F. A.: Reactive Oligomer-Based Plastic Foams. Moscow: Khimiya Press 1978 (Russ.)
89. Lowe, A. J. et al.: Proc. 4th Intern. SPI Conf. Cell. Plast. November 15–18, Montreal Canada, 1976, pp. 34; Chimica e l'Industria *61*, 656 (1979)
90. Schaver, A., Truxa, K., Spitzer, Z.: Stav. Časopis *15*, 245 (1967)
91. Lukas, W., Laskowski, W.: Polymer *16*, 281 (1971)
92. Guriev, V. V., Sinchillo, Yu. Ya., Shutov, F. A.: Plastich. Massy *5*, 41 (1978) (Russ.)
93. Shutov, F. A., Chaikin, I. I.: Pererab. Plastich. Massy *4*, 14 (1979) (Russ.)
94. Shutov, F. A., Chaikin, I. I.: Pererab. Plastich. Massy *6*, 17 (1979) (Russ.)
95. Guenther, F. O.: SPE Trans. *July*, 243 (1962)
96. Skochdopole, R. E.: Foamed Organic, CEP Technical Manual. New York: Reinhold Publ. Corp. 1961, pp. 21
97. Rubens, L. C., Skochdopole, R. E.: J. Cell. Plast. *1*, 52 (1965)
98. Saunders, J. H.: Rubber Chem. Tech. *33*, 1293 (1960)
99. Sandridge, R. L., Gemeinhardt, R. G.: Proc. ASTM Conf. 1961, pp. 31
100. Benning, C. J.: J. Cell. Plast. *5*, 48 (1969)
101. Deryagin, V. V. et al.: New Physico-Chemical Investigation Methods, in: Transactions of IFKh AN SSSR. Moscow: IFKh AN SSSR Publication 1957, pp. 2 (Russ.)
102. Briscall, H.: AWRE Report No 0-32/65. Berkshire U.K. 1965
103. Testerman, F., Hartman, C.: J. Appl. Phys. *34*, 1491 (1963)
104. Shutov, F. A.: A Previously Unknown Morphological Type of Phenolic Foam Structure, in: New Plastic Foam Technology, Vladimir: VNIISS 1974, pp. 90 (Russ.)
105. Fedoseev, V. I., Litvinova, G. A.: Koll. Zh. *38*, 756 (1976) (Russ.)
106. Berlin, A. A., Aseeva, R. M., Shutov, F. A.: Plastich. Massy *11*, 40 (1971) (Russ.)
107. Sceil, E.: Z. Anorg. Allg. Chem. *201*, 259 (1931)
108. Howard, R. T., Cohen, M.: Trans. AIME *172*, 143 (1947)
109. Smith, C. S., Guttmen, L.: J. Met. *4*, 150 (1952)
110. Johnson, W. A.: Met. Progr. *49*, 89 (1946)
111. Fullman, R. L.: J. Met. *5*, 447 (1953)

112. Reid, W.: J. Math. Phys. *34*, 95 (1955)
113. Gellings, R. J.: Appl. Sci. Res. *10*, 165 (1961)
114. Khodakov, G. S.: Small-Particle Physics. Moscow: Nauka Press 1972, pp. 33 (Russ.)
115. Lenz, F.: Z. Wiss. Microscop. *63*, 50 (1956/1958)
116. Harding, R. H.: Mod. Plast. *37*, 156 (1960)
117. Schael, G. W.: J. Appl. Polym. Sci. *11*, 2131 (1967)
118. Sobiczewski, Z.: Plaste, Kautschuk *10*, 735 (1963)
119. Lashnev, V. I.: Candidate Thesis. Kalinin: KPI Publication 1969 (Russ.)
120. Collins, F. H., Kraus, D. A.: SPE Techn. Papers *19*, 639 (1973)
121. Schwaber, D. M.: Polym. Plast. Technol. *2*, 231 (1973)
122. Menges, G., Knipschild, F.: Polym. Eng. Sci. *15*, 623 (1975)
123. Ollivier, J. P.: Rev. Gen. Caout. Plastiq. *53*, 65 (1976)
124. Dementiev, A. G., Tarakanov, O. G.: Mekhanika Polimerov *2*, 2258 (1972) (Russ.)
125. Dobrovolsky, I. P. et al.: Mekhanika Polimerov *1*, 151 (1971) (Russ.)
125a. Holte, K. G., Findley, W.: Trans. ASME, Ser. D, J. Basic Engng. *1*, 115 (1970)
126. Lohmeyer, S.: Gummi, Asbest, Kunststoffe *26*, 106 (1973)
127. Cherepanov, V. P., Shamov, I. V.: Plastich. Massy *10*, 53 (1974) (Russ.)
128. Chudnovsky, A. F.: Thermophysical Characteristics of Disperse Materials. Moscow: GIFML
     Publication 1962, pp. 456 (Russ.)
129. Novichenok, L. N., Shulman, Z. P.: Thermophysical Properties of Polymers. Minsk: Nauka
     i Tekhnika Press 1971, pp. 117 (Russ.)
130. Harding, R. H., James, B. E.: Mod. Plast. *39*, 133 (1962)
131. Ball, G. M., Hurd, R., Walker, M. G.: J. Cell. Plast. *6*, 66 (1970)
132. Kahlenberg, F.: Mod. Refrig. *67*, 391 (1964)
133. Hocking, C. S.: Plastvärlden *14*, 936 (1964)
134. Hayashi, F.: Kagaku go Koge *38*, 600 (1964)
135. Huldy, H. J.: Plastica *21*, 368 (1968)
136. Küster, W.: Kunststoffe *60*, 249 (1970)
137. Kadar, K.: Müanyanges gumi *10*, (1973)
138. Norton, F. J.: J. Cell. Plast. *3*, 23 (1967)
139. Patten, G. A., Skochdopole, R. E.: Mod. Plast. *39*, 149 (1962)
140. Skochdopole, R. E.: Chem. Progr. *57*, 55 (1961)
141. Tomanovskaya, V. F., Kolotova, V. E.: Freons. Moscow: Khimiya Press 1970 (Russ.)
142. Zehender, H.: Kühlt.-Klimat. *19*, 2 M (1967); Kunstst. im Bau No. 1, 3 (1979); Sonderdruck
     aus Isolierung No. 5, 3 (1980)
143. Schmidt, D. W.: J. Appl. Plastics *11*, 19 (1962)
144. Fischer, F.: Kunststoffe *56*, 321 (1966)
145. McIntire, O. R., Kennedy, R. N.: Chem. Eng. Progr. *44*, 727 (1968)
146. Jones, R. E., Patten, G., Steingiser, S.: Handbook of Foamed Plastics, New York: Lake
     Publ. Corp. 1965, pp. 52
147. Mittash, H.: Plaste und Kautschuk *16*, 268 (1969)
148. Just, M., Schöter, H., Eichenberg, C.: IFZ-Mitteilungen *11*, 225 (1972)
149. Shiina, N., Tsuchiga, M., Nakee, N.: Japan Plastic Age *10*, 37 (1972); *11*, 49; 45; 47 (1973)
150. Shutov, F. A.: A New Approach to the Relation Between Gas-Filled Polymer Structure and
     Properties, in: New Methods for the Production of Gas-Filled Polymers. Vladimir: BNIISS
     Publication 1978 pp. 17 (Russ.)
151. Berlin, A. A., Tsukerman, A. M., Shutov, F. A.: Plastich. Massy *10*, 18 (1979) (Russ.)
152. Shutov, F. A., Chaikin, I. I.: Pererab. Plastich. Massy *9*, 14 (1979) (Russ.)
153. Berlin, A. A., Shutov, F. A.: Strengthened Gas-Filled Plastics. Moscow: Khimiya 1980 (Russ.)
154. Berlin, A. A., Shutov, F. A.: Morphology of Oligomeric Plastic Foams, in: Physico-Chemistry
     and Chemistry of Reactive Oligomers. Alma-Ata: AN Kaz. SSR Publication 1979, pp. 50
     (Russ.)
155. Shutov, F. A.: Third Intern. Conf. on Application to Plastics in Industry, Plovdiv Bulgaria,
     1981 (Russ.)
156. Shutov, F. A.: Gordon Reseach Conf. on Foams, Plymouth USA, 1982

G. and S. Olivé (editors)
Received June 18, 1982

Author Index Volumes 1–51

Subject Index

CATALYSIS

Science and Technology

Editors: J. R. Anderson, M. Boudart

Volume 4

1983. 106 figures. X, 291 pages. ISBN 3-540-11855-1

Chapter 1, *P. N. Rylander:* **Catalytic Processes in Organic Conversions:** Introduction. – Hydrogenation. – Dehydrogenation.
Chapter 2, *H.-P. Boehm, H. Knözinger:* **Nature and Estimation of Functional Groups on Solid Surfaces:** Introduction. – Characterization, Identification, and Estimation. – Generation of Surface Groups.
Chapter 3, *G. Ertl:* **Kinetics of Chemical Processes on Well-defined Surfaces:** Introduction. – Kinetic Concepts. – Model Studies on Simple Catalytic Reactions. – Conclusions.

Volume 3

1982. 91 figures. X, 289 pages. ISBN 3-540-11634-6

Chapter 1, *E. E. Donath:* **History of Catalysis in Coal Liquefaction:** Introduction. – Historic Outline of the Bergius-Pier Process. – Primary Coal Liquefaction. – Refining of Coal Oils.
Chapter 2, *G. K. Boreskov:* **Catalytic Activation of Dioxygen:** Introduction. – Interaction of Dioxygen with the Surface of Solid Catalysts. – Isotopic Exchange between Dioxygen Molecules. – Oxydation of Dihydrogen. – Oxidation of Carbon Monoxide. – Catalytic Oxidation of Hydrocarbons and other Organic Compounds. – Interaction of Dioxygen with Sulfur Dioxide.
Chapter 3, *M. A. Vannice:* **Catalytic Activation of Carbon Monoxide on Metal Surfaces.** – Introduction. – The CO Molecule and Chemisorption. – CO Adsorbed on Metal Surfaces. – CO-H_2 Co-Adsorption and Interaction on Surfaces. – Methanation and Fischer-Tropsch Reactions. – Water Gas Shift Reaction. – Summary.
Chapter 4, *R. Morrison:* **Chemisorption on Nonmetallic Surfaces:** Introduction. – The Adsorption of Water. – Adsorption on Acidic and Basic Sites. – Adsorption by Local Bonding: Surface States. – Ionosorption. – The Effect of Adsorption on the Properties of the Solid.
Chapter 5, *Z. Knor:* **Chemisorption of Dihydrogen:** Introduction. – Metals. – Nonmetals. – Concluding Remarks.

Springer-Verlag Berlin Heidelberg New York Tokyo

Volume 2

1981. 145 figures. X, 282 pages. ISBN 3-540-10593-X

Volume 1

1981. 107 figures. X, 309 pages.

Springer-Verlag
Berlin
Heidelberg
New York
Tokyo